The Biology of Polar Habitats

Biology of Habitats

Series editors: M. Crawley, C. Little, T. R. E. Southwood, and S. Ulfstrand.

The intention is to publish attractive texts giving an integrated overview of the design, physiology, ecology, and behaviour of the organisms in given habitats. Each book will provide information about the habitat and the types of organisms present, on practical aspects of working within the habitats and the sorts of studies which are possible, and will include a discussion of biodiversity and conservation needs. The series is intended for naturalists, students studying biological or environmental sciences, those beginning independent research, and biologists embarking on research in a new habitat.

The Biology of Rocky Shores
Colin Little and J. A. Kitching

The Biology of Polar Habitats
G. E. Fogg

The Biology of Ponds and Lakes
Christer Brönmark and Lars-Anders Hansson

The Biology of Streams and Rivers
Paul S. Giller and Bjorn Malmqvist

The Biology of Polar Habitats

G. E. Fogg

School of Ocean Sciences,
University of Wales, Bangor

OXFORD NEW YORK TOKYO

Oxford University Press

1998

Oxford University Press, Great Clarendon Street, Oxford OX2 6DP

Oxford New York
Athens Auckland Bangkok Bogota Bombay Buenos Aires Calcutta
Cape Town Chennai Dar es Salaam Delhi Florence Hong Kong Istanbul
Karachi Kuala Lumpur Madrid Melbourne Mexico City Mumbai
Nairobi Paris São Paolo Singapore Taipei Tokyo
Toronto Warsaw

and associated companies in
Berlin Ibadan

Oxford is a trade mark of Oxford University Press

Published in the United States
by Oxford University Press, Inc., New York

A catalogue record for this book is available from the British Library

Library of Congress Cataloging in Publication Data

Fogg, G. E. (Gordon Elliott), 1919–
The biology of polar habitats / G. E. Fogg.
(Biology of habitats)
Includes bibliographical references and index.
1. Ecology – Polar regions. 2. Natural history – Polar regions.
I. Series.
QH541.5.P6F64 1998 578′.0911 – dc21 98-14970

ISBN 0 19 854954 7 (Hbk)
0 19 854953 9 (Pbk)

Typeset by Best-set Typesetter Ltd., Hong Kong

Printed in Great Britain by
Bookcraft (Bath) Ltd
Midsomer Norton, Avon

To
the Memory of
Elizabeth Beryl Fogg
who once ventured with me
into Antarctica

Preface

A great attraction in studying polar habitats is that, although it may involve sophisticated biochemical or electronic techniques, it still takes one into wild, physically challenging, and hauntingly beautiful places. Fortunately for those who want to indulge themselves in this way there are sound scientific reasons to support their applications for the necessary funding. Apart from the interest in finding out how living organisms manage to exist under the apparently adverse conditions, polar habitats provide particularly favourable opportunities for investigating basic ecological relationships. Compared with the communities of temperate and tropical regions, those in polar habitats are of recent origin and, as a result, relatively simple with few species. This makes it somewhat easier to identify the critical factors operating in the environment to determine the presence and abundance of individual species, interrelations between species, cycling of nutrients, and energy flow. A further feature adding to the interest is that although the Arctic and Antarctic are both cold, with nights and days which last for months, they have inverse patterns of distribution of land and sea. This makes for differences in climate and ocean circulation which have had profound effects on the invasion of plants and animals from temperate zones so that the respective ecologies of the two regions are different. Not unrelated to this, there are radical differences in political regimes which have resulted in the support and organization of science in the Arctic and Antarctic being remarkably dissimilar.

Knowledge of polar habitats is valuable from the practical as well as the academic point of view. The polar regions, taken as including the sub-Arctic and sub-Antarctic as well as the high Arctic and Antarctic, occupy a great part of the earth's surface and with increasing pressure from human populations it is advisable to know what resources and potentialities these relatively unused lands and seas might provide. Proper management of polar fisheries, mineral exploitation, oil drilling, and human settlement all depend on an understanding of the ways in which habitats work. Tourism, also, burgeoning in both regions, needs an input of this understanding if visitors are to get the best out of the experience with the least damage to the environment. Beyond this, polar ecology has world-wide significance. It is increasingly evident that polar ecosystems intermesh into global processes and that they play key roles in the regulation of the environment which is vital for the well-being of all mankind.

My principal object is to provide a readable account of polar habitats which will be useful to first degree students in biology – especially plant, animal, and microbial ecologists – in geography, and in environmental sciences. It may also serve as an introduction for those beginning research on polar topics and as a background for conservationists and, one hopes, the enquiring tourist.

I am grateful to many people for help with information and suggestions. These include Professor S. S. Abyzov, Professor T. V. Callaghan, Professor P. M. Jónasson, Dr J. Kain, the staff of the Museum of Mankind, London, Dr B. Stonehouse, Professor E. Sunderland, Dr D. W. H. Walton, Dr C. Wiencke, and, especially, Dr David Jones, who put the idea of a book such as this into my head, Professor Andrew Clarke, for his constructive review of the manuscript, and my editor, Dr C. Kennedy, for her general helpfulness. My thanks are also due to those, to whom acknowledgement is made in appropriate places in the text, who have generously allowed the use of illustrations. Finally, I must thank my wife for her tolerance of my undoubtedly exasperating preoccupation with yet another book.

Menai Bridge, Isle of Anglesey, April 1997 G. E. Fogg

Contents

1 The physical background

1.1 Introduction

The severities and complexities of polar habitats stem from simple geometry: the axis of the earth's rotation is not at right angles to the plane in which it orbits round the sun. As a result, the North and South Poles are tilted in turn towards the sun and, instead of having the daily alternation of day and night to which most of us are accustomed, have months of nightless summer followed by sunless winter. With all his ability to create his own environment man does not find it easy to cope with either midnight sun or a night of several months. To a greater or lesser extent other organisms also have problems in adapting to the polar night. Surprisingly, the frigid climate which we tend to think of as the most characteristic and hostile feature of the poles is not a necessary result of the smaller amount of solar radiation which they receive but has been brought about by interplay of secondary factors. It has not been the norm in the geological past. The varying input of radiant energy from the sun over the surface of the earth sets up circulations in the atmosphere and oceans, which, directed by forces produced by the earth's rotation and by the positions of land masses, convey heat from one part to another. In the immediate geological past these factors have operated to cause ice to accumulate at the poles. At present, ice is a dominant feature in polar habitats and its manifold patterns of formation, structure, accumulation, and movement give rise to a wide variety of situations, some transient, others more lasting, in which organisms have established themselves. In some, conditions are so severe that life survives only in a dormant state. Others support a surprising variety, amount, and activity of living things. The nature of the environmental conditions in these habitats and the ways in which organisms adapt to them provide the subject matter of this book. However, before considering these biological aspects we must look at the general physical features of the polar regions.

One definition of 'polar regions' might be that they are the areas within the Arctic and Antarctic Circles. These are parallels of latitude at 66°33′ north and south, respectively, corresponding to the angle between the axis of rotation of the earth and the plane of its orbit round the sun. These areas total some 84 million km^2, 16.5%, of the surface of the earth. The polar circles, though, mark no sharp transitions, either in climate or in flora and fauna, and they have little ecological significance. In places inside the Arctic Circle there are forests and thriving towns. Within the Antarctic Circle there is nothing but sea, ice, and sparse exposures of rock at the present day (Fig. 1.1). However, 60 million years ago, in the early Cenozoic, forests flourished despite Antarctic winter darkness. The polar regions are defined

(a)

(b)

Fig. 1.1 Contrasting climates north and south in situations inside the polar circles and on comparable west-facing sheltered coasts at around midsummer: (a) Tromsø, 69°42′N 19°00′E; and (b) Stonington Island, 68°12′S 67°00′W. (Photos G. E. Fogg.)

in various other ways by climatologists, terrestrial ecologists, marine biologists, geographers, and lawyers, but these will be considered later.

1.2 The energy balances of the polar regions

1.2.1 Solar irradiance

About half the solar energy entering the atmosphere consists of visible radiation, most of the rest is infrared, and a small fraction is ultraviolet. Mean values for total direct radiation from the sun penetrating to the earth's surface at various latitudes in the northern hemisphere are shown in Fig. 1.2. Because they have the sun for all or most of the 24 hours, polar situations actually receive more radiation around midsummar than do those on the equator. Nevertheless, the low angular height of the sun, even at midsummer, and its disappearance in winter result in the total radiation per unit surface delivered during the year at the North Pole being less, about 43%, than that at the equator. Direct solar radiation is augmented by scattered light from the sky to a variable extent, usually about 20% of the total. When the sun is obscured by cloud much of its radiation is reflected back into space so that the values in Fig. 1.2, although they do not include scattered radiation from the sky, are likely to be overestimates.

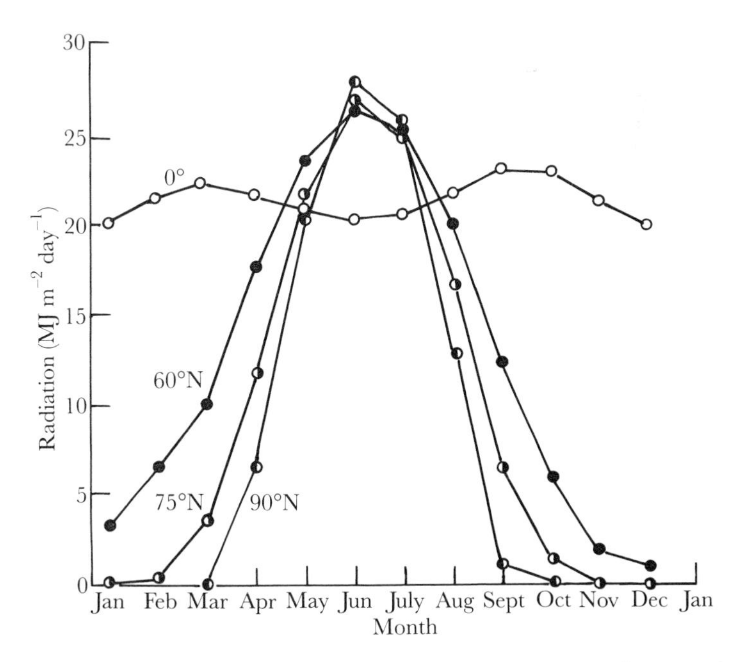

Fig. 1.2 Estimates of total direct radiation, with corrections for variations in atmospheric turbidity, on the 15th day of each month at sea level at various latitudes north. (Data of Perl 1935 from Hutchinson 1957.)

Table 1.1 Albedos of various natural surfaces

Snow-covered sea ice	0.95
Fresh snow	0.8–0.85
Melting snow	0.3–0.65
Quartz sand	0.35
Granite	0.15
Bare earth	0.02–0.18
Coniferous forest	0.10–0.14
Water	0.02
The earth as a whole	0.43

These generalizations apply to both polar regions but there are differences. Since the earth is closest to the sun in the austral summer but most distant in the boreal summer, 7% more energy enters the Antarctic than the Arctic. Furthermore, the Antarctic atmosphere has less radiation-absorbing dust and pollutants, and, the continent having a higher elevation, the atmospheric mass to be penetrated by incoming radiation is less. Together, these factors result in the Antarctic getting 16% more energy. Nevertheless, the Antarctic is the colder. The reasons for this will become apparent in the following sections.

1.2.2 Reflection and absorption of solar radiation

Incident radiation falling on a body may be reflected, transmitted, or absorbed. Absorbed radiation is changed into thermal energy in the absorbing material. The ratio of reflected to incident radiation is known as the *albedo* and has the values of 1.0 for complete reflection and 0 for complete absorption. Snow and ice have high albedos, water a low albedo, and rocks are intermediate (Table 1.1). The mean albedos of the polar regions vary seasonally but are always higher than that of the earth as a whole. That of the Arctic is lower than that of the Antarctic, 0.65 as compared with 0.90, as a result of loss of reflective snow cover and relatively greater ice melt in summer. In the Arctic ice covers about 2 million km^2 of land and sea ice extends over 7 million km^2 at its minimum and 14–16 million km^2 at its maximum in late February or March. In the Antarctic, land ice extends over 12.6 million km^2 and sea ice over 4 million km^2 at its minimum increasing to about 20 million km^2 at its maximum in September, when it goes well north of the Antarctic Circle all round the continent and effectively more than doubling its size (Fig. 1.3a,b). The total area of high albedo in the summer is sufficient at both poles to reflect much of the incident radiation back into space and thus reduce heating of land and sea. This resolves the paradox that the area which receives the maximum monthly input of solar energy of any on earth, the ice sheet of East Antarctica, is also the coldest on earth.

Spots of low albedo within polar regions can absorb large amounts of heat. The Russian station, Mirny (66°33′S 93°01′E),* is on snow-covered ground whereas the nearby Oazis station (66°30′S 101°E) has bare rock around it. At Mirny, most of the incoming radiation is reflected and little heat is accumulated in the ground even at midsummer – because of its high albedo, snow cover tends to persist once established. At Oazis, the rock surface heats up in summer and soil temperatures rise to 10°C or so above that of the ambient air. A water surface, which also has a low albedo except at low angles of incidence, behaves similarly. Since water has a high specific heat, lakes and seas act as particularly effective heat stores. Not only does water transmit radiation into its depths but heat can be carried downwards by its turbulence. Were it not for extensive and persistent high-albedo snow cover, polar climates could be temperate, as, indeed, they have been in the past.

1.2.3 Long-wave radiation from terrestrial sources and its absorption in the atmosphere

Thermal energy acquired by absorption of solar radiation is lost by emission of radiation of longer wavelength – infrared. The amount of energy reradiated is a function of the infrared emission characteristics of the surface and the fourth power of its absolute temperature. Most natural surfaces have emission characteristics in the same range – snow, ice, rock, and water all having similar high values at the same temperature. The earth's surface, to a good approximation, can be regarded as having perfect infrared emissivity at a temperature of 285 K in the waveband 4.5 to 50 μm, with a peak at about 10 μm.

Whereas the atmosphere is highly transparent to solar radiation it absorbs terrestrial radiation because of the presence in it of clouds, water vapour, and certain gases, all of which show high absorption within the waveband just specified. These gases, which include carbon dioxide and methane, have achieved notoriety as 'greenhouse gases' because their increasing concentrations in the atmosphere, leading to increasing interception of infrared radiation, have given rise to fear of global warming. Without the blanketing effect of the atmosphere the earth's surface temperature would be 30 to 40°C lower than it is and would vary between greater extremes of heat and cold. Just now the blanket is becoming oppressively thick.

Liquid water in the form of clouds is nearly opaque to terrestrial radiation even though its concentration may be only $1\,g\,m^{-3}$ – equivalent to a thickness of $0.001\,mm\,m^{-1}$. Clouds of ice or snow are similarly highly absorbing.

Water vapour present in the clear atmosphere also has high absorption for most wavelengths in the terrestrial emission spectrum but has a window in the region of 10 μm, the region of maximum terrestrial emission. The frost

*The latitude and longitude of places are given on first mention in the text.

which often accompanies a cloudless night is a familiar example of the heat loss which this allows.

'Greenhouse gases' trap about 40% of the radiation from the earth's surface. Carbon dioxide is important because its absorption peaks fall in between those of water vapour, although they do not completely fill the gaps. The increase in its concentration in the atmosphere during the last hundred years as a result of burning of fossil fuels is a dominant factor causing the rise in global temperature. At its present level of about 350 ppm (parts per million) carbon dioxide absorbs most radiation within the regions of its absorption peaks. Gases having absorption peaks filling other gaps are now becoming relatively more important. Thus, an increase in methane concentration has 25 times the effect of a similar increase in carbon dioxide and for chlorofluorocarbons (CFCs) the effect is ten thousand times greater.

1.2.4 Long-wave radiation emission in the atmosphere

The absorbing agents in the atmosphere heat up and in their turn emit long-wave radiation according to their temperature and emission characteristics. Some of this, about 75%, will return to the earth's surface and there be reabsorbed. The net loss of energy from this surface will thus be that of the radiation which it emits itself less that of the back radiation received from above. When a layer of warm cloud overlies cold ground the balance becomes positive. The rest of the long-wave radiation emitted by atmospheric components escapes into space. There is an overall heat loss via long-wave radiation from the earth and its atmosphere because their temperatures are higher than that in space.

Net heat loss is greater in the Antarctic than in the Arctic. The reason for this lies partly in the greater prevalence of clouds in the Arctic, particularly at its periphery, as compared with the Antarctic, the high continental plateau of which is generally cloud-free. Furthermore, the atmosphere over the plateau is more transparent to long-wave radiation because of its thinness and dryness. A complication in both polar regions is added by temperature inversions, that is to say increases, rather than decreases as normally found, of temperature with height above ground. These arise as a result of snow surfaces beneath clear skies reflecting nearly all incoming radiation, so that air near the surface becomes chilled and dense. Above it, at 200–1000 m, is less dense air containing more moisture which intercepts some outgoing long-wave radiation and keeps warmer. Inversions are prevalent over the Antarctic plateau for most of the year and account for the fact that temperatures on the plateau, after a rapid fall in autumn, scarcely decrease thereafter so that the winter is 'coreless' (Fig. 1.6). Variations in refraction associated with inversions produce the optical phenomena, such as mirages, which are characteristic of polar regions (Pielou 1994).

1.2.5 Transport and global balance of thermal energy

Around the poles there is net loss of energy by radiation over the year whereas in equatorial latitudes there is a net gain. Losses balance gains and as a whole the earth and its atmosphere neither warm up nor cool. The loss of heat from the polar regions is made good by a flow of excess heat carried in currents of air or water from lower latitudes. The poles are sinks for thermal energy and the equatorial regions the source. Air transports not only sensible heat but also latent heat in the water vapour it contains. Some is released when the vapour condenses to form clouds (539 cal g^{-1}) and some more when the liquid water freezes (79.8 cal g^{-1}).

If Ptolemy's geocentric theory were correct and the earth remained motionless while the sun went round it, warm air carrying water vapour with it would rise in equatorial regions and flow towards each pole along meridional paths. Cooling on the way it would eventually sink and return towards the equator, again along meridional tracks, at a lower level. In the oceans, seawater, concentrated by evaporation in low latitudes would become saltier, more dense, and sink, likewise flowing polewards carrying heat along meridional pathways assuming no obstruction by land masses. In high latitudes it would rise as it encountered water which, although less salty, was denser because it was colder. However, because the earth rotates, flows of air and water are deflected, by the Coriolis force, to the right of the direction of movement in the northern, and to the left in the southern hemisphere. The Coriolis force is zero at the equator and maximal at the poles.

In the atmosphere, thermal energy is mostly transported in the lower layer, the *troposphere*, which is about 10 km thick and separated by a temperature minimum, the *tropopause*, from the *stratosphere*. The basically meridional two-way traffic of warm air polewards and cool air equatorwards is obscured, not only by the earth's rotation, but also by the different thermal effects of continents and oceans. The result is a complex pattern of zonal and cellular circulation. Salient features are that, between latitudes 30° and 60°, both north and south, there are zones of predominantly low pressure and westerly winds, and polewards of 60° there are zones of high pressure with northeasterly and southeasterly winds, respectively, north and south.

Water currents are also subject to the Coriolis force but are obstructed and deflected by land masses and irregularities in the seabed. A further complication is that atmosphere and ocean interact. The drag of winds on the sea surface induces currents and, also, by setting up slopes in the surface, winds produce other currents in response to the pressure gradients which arise. Wind-induced currents may be temporary, varying with local weather but the major oceanic circulations correspond roughly to the pattern of the prevailing winds. Such currents are largely superficial, and, for present purposes, deep water currents, the direction of which need show no relation to those at the surface, are more important. The temperature at great depths in the oceans is everywhere near to freezing point. This cold bottom water comes from two main sources, one in the Greenland Sea, the other in the Weddell Sea in the Antarctic, where surface water becomes cold and dense enough to sink to the bottom and flow equatorwards. Other deep water currents carry warm salty water from equatorial regions polewards in replacement. These currents are of enormous volume but move slowly, perhaps around one or two kilometres per month, and so the water in them stays below the surface for many hundreds of years.

1.2.6.1 Heat influx and balance in the polar regions

Against this general background we can look more specifically at the paths by which thermal energy reaches the polar regions. First, a radical geographic difference between Arctic and Antarctic is to be noted. Whereas the Arctic centres on a sea of some 14 million km^2 enclosed by islands and the northern stretches of continents, the other is a continent of 13.3 million km^2 – larger than Europe but smaller than South America – surrounded by a belt of ocean which separates it by 1000 km from an outlier of the nearest major land mass (Fig. 1.3). This difference has profound consequences for their respective climates, biology, and importance in the regulation of the global environment (Walton 1987).

1.2.6.2 The Arctic

The major input of heat is provided by northward moving warm air which interchanges with cold polar air in cyclones associated with low pressure along the atmospheric polar front in the region of 60°N. Variations in surface topography introduce complications and a regular succession of cyclones is frequently obstructed by well-developed stationary regions of high pressure – 'anticyclonic blocking'. From there, the warm air travels high in the troposphere to subside around the pole. It then returns as surface winds away from the pole, the Coriolis force giving these an easterly direction. This is an anticyclonic situation. Over the sea ice of the Arctic Ocean the motion of these winds is imparted by pressure differences. On the Greenland ice cap, winds become related to topography, dense cold air flowing downslope. Such *katabatic* winds are intermittent. The air accelerates as it descends, becoming compressed by the higher pressure at the lower levels and developing heat equivalent to the work done.

Heat is also contributed by the great Siberian rivers, which introduce about 3500 $km^3 yr^{-1}$ of freshwater, mainly in summer when its temperature may get up to 10–15 °C. An additional 1500–2000 $km^3 yr^{-1}$ enters as a freshwater fraction of the Bering Strait inflow. The main ocean current flowing into the Arctic Ocean (Carmack in Smith 1990) is the West Spitsbergen Current, a northward-flowing extension of the Norwegian Atlantic Current, passing through the Fram Strait (*c.* 80°N 0°, Fig. 1.4). This follows a deep trench leading to the Arctic Ocean. The apparent access via the Barents Sea is obstructed by shallows. The water in this current is warm (above 3 °C) and relatively saline (greater than 34.9‰). The amount of water transported is uncertain, estimates vary between 2 and 8 Sv (1 Severdrup = $10^6\,m^3\,s^{-1}$), or 60 000 and 250 000 $km^3\,yr^{-1}$, but it is possible that as much as half of this

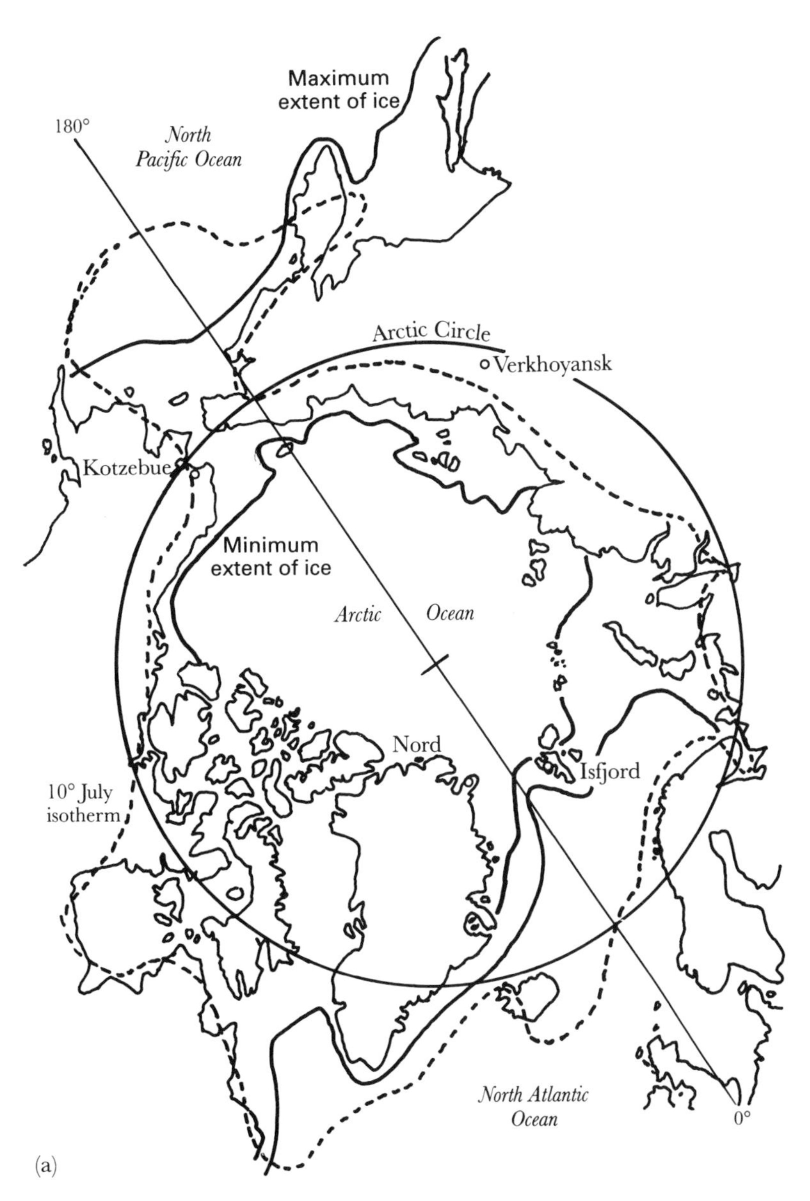

Fig. 1.3 Extents of sea ice: (a) Arctic (with the weather stations as in Fig. 1.6 and 10 °C July isotherm); and (b) Antarctic (with the weather stations as in Fig. 1.6 and some major research stations).

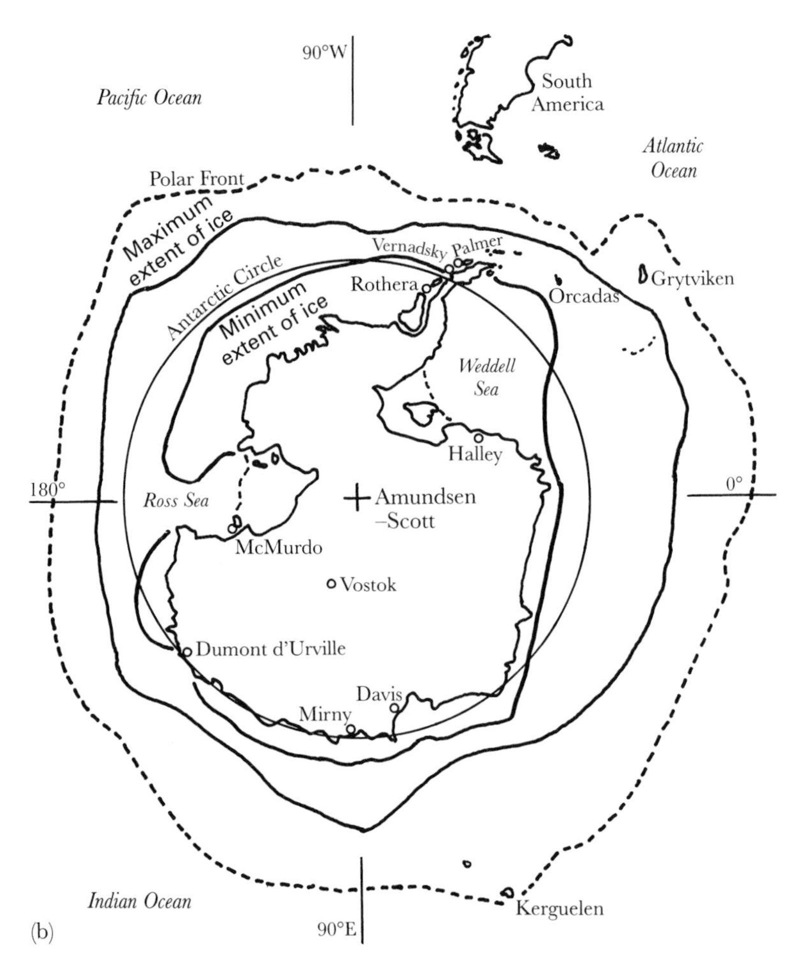

Fig. 1.3 Continued

circulates in the vicinity of the Fram Strait without entering the Arctic Ocean. Alongside the West Spitsbergen Current to the west is the East Greenland Current. This is the main current out of the Arctic Ocean and carries cold (below 0 °C), relatively fresh water (less than 34.4‰ salinity), south. The flow of this current is between 3 and 30 Sv (91 000 to 910 000 $km^3 yr^{-1}$) and it carries with it some 4 million megatons of drift ice to lower latitudes. The Bering Strait is narrow (85 km) and shallow (50 m), allowing a small (about 0.8 Sv or 25 000 $km^3 yr^{-1}$) northerly flow. Within the Arctic Ocean the Transpolar Drift, a surface current, flows from the Siberian to the Greenland side, where it feeds into the East Greenland Current. It was trusted by Nansen to carry his ship, the *Fram*, beset in the ice, across the Arctic Ocean into the vicinity of the North Pole. A less heroic demonstration of the Transpolar Drift is being provided (1996) by a consignment of yellow plastic ducks, lost from a container ship in the North Pacific at 44°N 178°E in October 1992. The ducks are expected in the North Atlantic

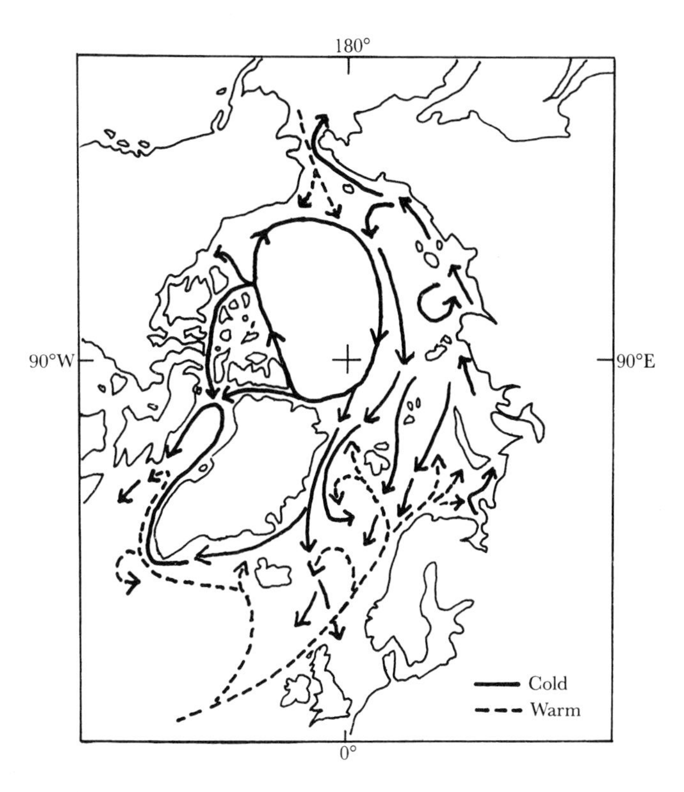

Fig. 1.4 Principal surface currents in the Arctic seas.

around the turn of the century. The pack ice, up to 3 m thick, plays an important part in conserving heat. The seawater itself, with a mean depth of about 1200 m and a volume of about 17 million km^3, provides an immense heat reservoir. Its ice cover reduces heat transfer to the atmosphere by one or two orders of magnitude compared with that from open water. Furthermore, largely because of the inflow of river and Bering Strait waters, a layer of low salinity floats on top of the denser water in the Arctic basin, producing a marked *halocline* at between 30 and 60 m which limits the convection which would otherwise mix the whole water column and promote heat loss. The deep water consequently remains between −0.5 °C and −0.9 °C, appreciably above its freezing point of −2.0 °C. These various factors contribute to the generally higher temperatures of the Arctic as compared with those of the Antarctic. Plans to divert southward part of the flow of some Siberian rivers to alleviate water shortages in the south perhaps need not create too much alarm. On present evidence it seems unlikely that such diversions would have major effects on circulation or sea ice distribution in the Arctic Ocean.

1.2.6.3 The Antarctic

Although the heat exchanges of Antarctica are still incompletely understood, it presents a simpler situation for analysis, having an approximately circular ice-covered land mass forming a dome, without too many topographical irregularities, nearly centred on the geographical pole and surrounded by a continuous belt of deep ocean (Walton 1987). This allows the collection of meaningful data and the construction of realistic mathematical models of circulatory processes.

The chief agent of imput of thermal energy is again atmospheric circulation. The zone of westerly winds produces a succession of cyclones which satellite images show as a regular procession of cloud spirals around the continent in latitudes 60–65°S. The Antarctic thus contrasts with the Arctic in that anticyclonic blocking is infrequent. The cyclones often swing south into the Ross Sea area but rarely depart from their circumpolar track to carry warm air into the centre of the continent. As in the Arctic, cyclones provide a mechanism for exchanging cold polar air for warm moist air from lower latitudes. The water vapour is again a source of thermal energy. This air travels south at an intermediate height in the troposphere and sinks in the high pressure region over the summit of the polar plateau. The ice dome favours the initiation of katabatic winds, which under the influence of the Coriolis force follow a northwesterly track. This layered system of air movements is seen in dramatic form when smoke from the volcano Erebus (3785 m, 77°40′S 167°20′E) is carried polewards whereas at sea level a blizzard may be blowing from the south. If these low level winds are sufficiently strong, surface irregularities such as wind-produced ridges in the snow (*sastrugi*) cause turbulence which disturbs inversions (p. 6) and mixes in warmer air and moisture from above. The speed of katabatic winds increases with slope and so turbulent heat exchange is about four times greater around the edge of the continent than it is in the interior.

There are no rivers to contribute to the heat budget of the Antarctic. In the sea (Carmack in Smith 1990) there is meridional transport which is, however, deflected by the wind-driven Antarctic Circumpolar Current. This current, being deep-reaching and confined only at the Drake Passage and south of the Australasian land mass, where it has to pass through a deep channel connecting the Indian and Pacific Oceans, is an enormous flow of about 130 Sv (4 million $km^3 yr^{-1}$). This is no mean barrier to meridional transport and is regarded as one of the major factors contributing to the exceptionally frigid state of Antarctica (see p. 28). Nevertheless, considerable southward transport of heat, mainly from the Indian Ocean, takes place in the Circumpolar Deep Water (Fig. 1.5), which has temperatures between 0 °C and 1.8 °C and maximum salinities around 34.76‰. This wells to the surface at the Antarctic Divergence, about 70°S, and spreads north and south. The north-flowing fraction mixes with freshwater released by ice melt, giving temperatures around −0.4 °C and salinities about 34.20‰. It forms a layer, some 200 m in depth, separated by a sharp density gradient (*pycnocline*) from the Circumpolar Deep Water. Meeting warmer Subantarctic Surface Water, it plunges beneath this at the Antarctic Polar Front, a feature also known by the not-quite-synonymous name of Antarctic

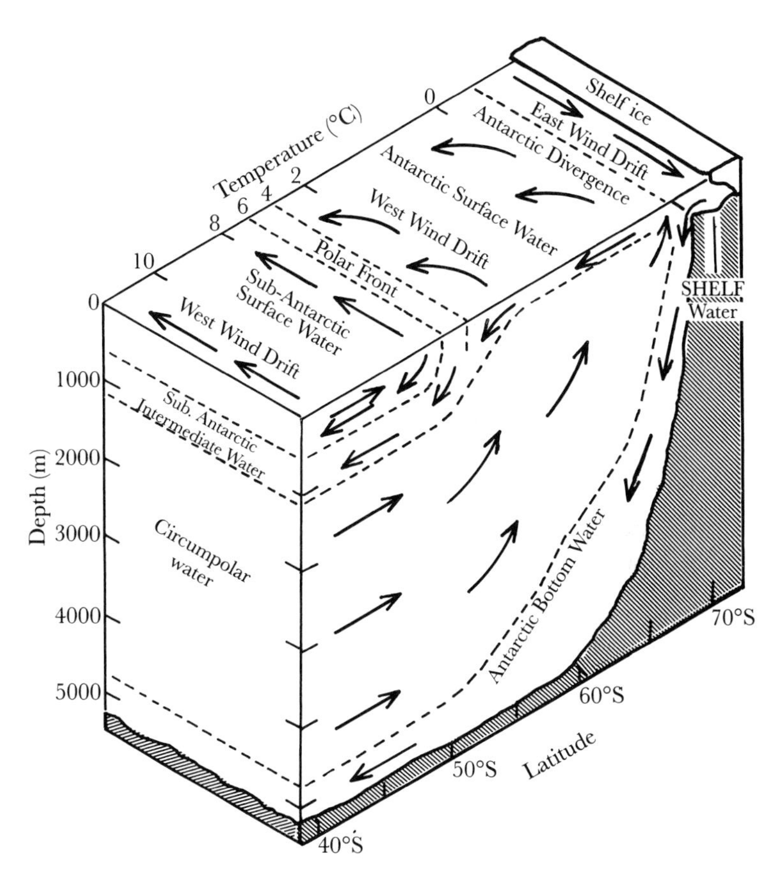

Fig. 1.5 Three-dimensional diagram showing the circulation patterns and water masses in the Southern Ocean.

Convergence (Fig. 1.5). The Polar Front remains in a surprisingly constant position, extending all the way round the continent within the Circumpolar Current (see Fig. 1.3b). It is a boundary of great biogeographical importance and is easily detected by abrupt changes in temperature in both surface seawater and air. The upwelling water from the Circumpolar Deep Water which continues on south becomes colder and mixes with water, from the continental ice shelves, charged with brine formed by freezing of seawater. Being both cold (0.4–1.3 °C) and saline (34.66–34.72‰) it sinks and then spreads northwards across the sea bottom. The Weddell Sea and to a lesser extent the Ross Sea are the major centres for the production of this Antarctic Bottom Water. Traces of it have been found as far beyond the equator as 17°N in the Atlantic. The scale of events in the Weddell Sea is shown by the estimate of between 76 and 97 Sv (2.4–2.0 million $km^3 yr^{-1}$) for water transport in the gyre occupying its basin. This vastly exceeds that in Arctic waters.

The area of the Southern ocean covered by sea ice increases five- or sixfold each winter (Fig. 1.3b) but with great year to year variation in timing, extent, and distribution. This interannual variability is linked to atmospheric processes and the flow rates and directions of ocean currents (p. 181; Murphy *et al.* 1995). Unlike Arctic ice, which is constrained by land, Antarctic ice is free to spread over deep ocean, almost anywhere that wind and tide take it. The heat exchanges of the Southern Ocean are correspondingly modified. Ice formation is most active in coastal regions subject to cold katabatic winds – further offshore turbulence retards ice formation. Both pack ice and icebergs are carried in a generally northern direction and, since freezing involved release of latent heat, whereas the eventual melting requires supply of heat, this implies a net poleward transport of heat against an export of freshwater. The release of heat by freezing at the beginning of winter and its uptake on melting in summer work to buffer temperatures. The high albedo and insulating properties of the ice also minimize heat exchanges. Leads (lanes) and patches of open water are perhaps more extensive in Antarctic sea ice than once supposed so the insulating effect may not be as great as in the Arctic Ocean.

The Arctic and Antarctic are the two great heat sinks which between them determine the patterns of both atmospheric and oceanic circulations and are thus key areas in regulating the global environment. Of the two, the Antarctic is the dominant.

1.3 Climate

1.3.1 The climatic boundaries of the polar regions

A simple definition of the polar regions is that they are those areas in the vicinity of the poles where the mean temperature of the warmest month is less than 10 °C. The 10 °C summer isotherm (Fig. 1.3a) usually coincides with the limits of tree growth (Aleksandrova 1980). Isolines of radiation balance give a better match to tree line although sometimes deviating by as much as 160 km. The position of the tree line depends on both latitude and altitude. The transition from tall forest to dwarf, shrubby, vegetation, which marks the tree line, is sometimes strikingly sharp, largely because single exposed trees, not being able to ameliorate their environment, tend to be eliminated. A closed canopy affords some protection from wind stress but since forest vegetation is penetrated by large-scale eddies temperatures of the above-ground tissues are closely coupled to those in the air. In contrast, shrubby vegetation is aerodynamically smoother and dissipates heat less readily, experiencing tissue temperatures and microclimates that, on average, are warmer than the air. Consequently, dwarf shrubs can succeed in polar climates in which trees fail to grow and reproduce.

In the Arctic the 10 °C summer isotherm undulates around the Polar Circle, going well south of it in the regions of the Bering Strait and the Northwest Atlantic where the Kamchatka and Labrador Currents, respectively, bring cold surface water down from the north. The isotherm goes north along the coast of Norway because of the warm North Atlantic Drift (Fig. 1.4). Around the Antarctic the 10 °C summer isotherm runs well north of the Polar Circle, at about 50°S, and almost entirely over ocean, only touching land at the tip of South America. There the coast is mostly forested. The sub-Antarctic islands south of the isotherm have vegetation which resembles Arctic tundra and are treeless. The Falkland Islands (*c.* 53°S 58°W) lie just on the cold side of the isotherm and have no native trees but their grasslands, dwarf shrub heaths, and *fell fields* (with discontinuous cover of cushion plants) are scarcely sub-Antarctic in character and will not be dealt with here.

1.3.2 The Arctic climates

Within the confines of the 10 °C summer isotherm conditions are generally cold, dry, and windy (Sugden 1982; Stonehouse 1989) but there are variations which are not easy to classify. The Arctic can be divided into the central maritime basin and the areas peripheral to it – in which can be distinguished the ice caps, polar maritime climates (located principally around the Atlantic and Pacific coastlines), and the polar continental climates as in north Alaska, Canada, and Siberia. There are no fixed meteorological stations in the central maritime basin but observations have been made from a succession of stations on drifting ice islands. This is a climatically stable area with a strong central anticyclone, clear skies, and light centrifugal winds during the winter. Because of the reservoir of heat in the ocean, temperatures do not fall to extremely low levels, averaging −30 °C offshore and −26 to 28 °C in the coastal regions during the depth of winter. When the sun returns the anticyclone weakens and there are incursions of depressions, bringing moist air, fog, cloud, snow, rain, and strengthening winds. Temperatures, except in a small central area, rise above freezing point so that the periphery of the pack ice melts and large areas of open water appear along the Alaskan and Eurasian coasts during June and July.

There are ice caps on the more northerly islands, for example, about 58% of Svalbard (*c.* 79°N 15°E) is permanently ice-covered. That on Greenland, which covers most of the island and rises to 3000 m, is by far the most massive. It is fed by snow borne by year-round southwestern airstreams, which deposit some 100 cm rain equivalents annually in the south but only about 20 cm in the north, parts of which are consequently almost ice-free. Temperatures on the plateau of the ice cap fall to −40 to −45 °C in winter, rising to −12 °C in summer.

Arctic continental climates are dominated by anticyclonic conditions in winter with low temperatures, light winds, and little precipitation. Mean monthly temperatures rise to freezing point around May and can get well above it in the short summer (Fig. 1.6). Weakening of the anticyclone allows incursions of depressions, bringing warm, moist, oceanic air with

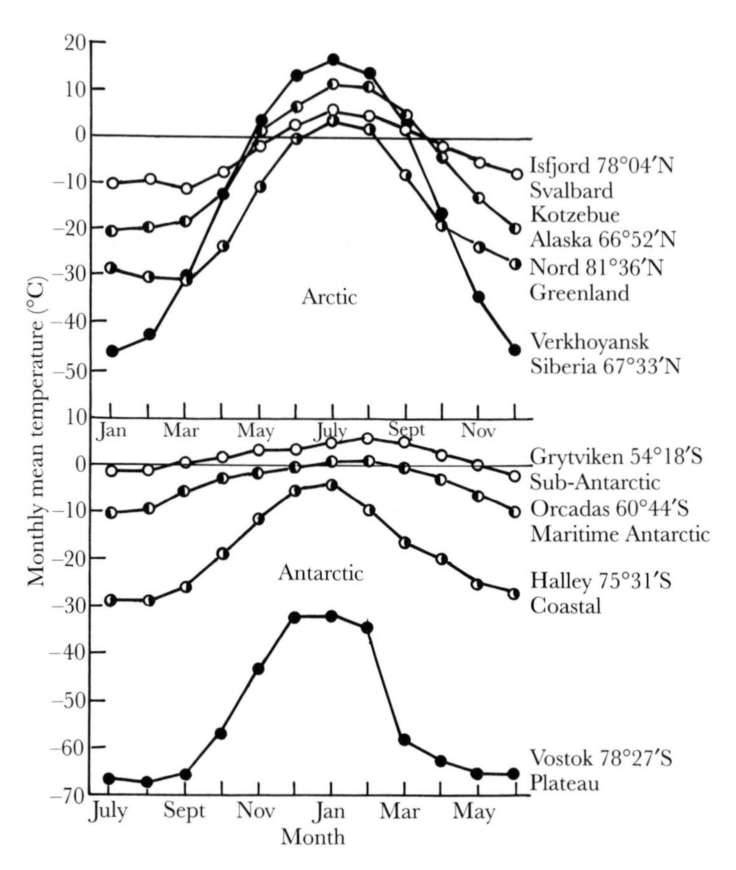

Fig. 1.6 Mean monthly temperatures at different locations (see Figs 1.3a,b) in the Arctic and Antarctic. Note that at Vostok the temperature from April to September remains more or less level (i.e. the winter is 'coreless', unlike that in the Arctic, which has a sharp minimum). (Data from Stonehouse 1989.)

precipitation that favours the development of tundra on low ground and ice caps and glaciers higher up. The most extreme continental conditions are found around Verkhoyansk (67°35′N 133°25′E), the 'pole of cold', in eastern Siberia, where an intense winter anticyclone spreads cold, dry, air in all directions. Being well away from the sea it has variations in temperature from −67.8 °C in winter to 36 °C in summer (Fig. 1.6). Precipitation is mainly in the form of summer rain and amounts to only 15 cm per annum.

Maritime climates are ameliorated by the sea, especially where there are warm currents. In the Canadian Arctic, the worst climate is encountered in the Hudson Strait area (*c.* 63°N 70°W), which is dominated by open water and frequent cyclonic activity, giving the highest average temperature, but the heaviest snowfall, highest average wind speeds, and greatest number of summer fogs in this sector. This is the region in which many of the early seekers after the Northwest Passage came to grief. The southwest of Greenland is warmed by an offshoot of the North Atlantic Drift (see Fig. 1.4) and is comparatively free of sea ice. Its mild climate allows sheep farming on luxuriant tundra within a short distance of the ice cap. Parts of Iceland likewise have a mild climate and forests of birch and spruce in the south. Its northern shores have pack ice drifted in by the East Greenland Current. The same current keeps the east coast of Greenland cold, even in summer.

The North Atlantic Drift passes Iceland to give Svalbard (Fig. 1.6) and the northwestern tip of Europe remarkably temperate climates with the tree line going far north. Only the northern part of Svalbard remains ice-bound in summer. The effect of the same current persists along the Eurasian coast of the Arctic Ocean as far east as Novaya Zemlaya (*c.* 75°N 60°E), keeping the Barents Sea open in summer. Depressions bring abundant summer rain as well as winter snow. The great Siberian rivers have some ameliorating influence but further east still the coastal climate becomes harsher with short, cold, summers and frigid winters. Precipitation decreases and Kotel'niy (75°59′N 138°00′E) on Ostrova Novosibirskiy, which has only 13 cm per year, can be described as desert.

1.3.3 The Antarctic climates

The array of Antarctic climates is simpler (Sugden 1982; Stonehouse 1989). The central feature here is the enormous, high, continental plateau, usually dominated by a high pressure system. With its 'coreless' winter (p. 6) goes a 'pointed' summer, lasting only a few weeks (Fig. 1.6). When planning their attempts on the South Pole, neither Amundsen nor Scott had any idea of this state of affairs. Amundsen was lucky to arrive just before the peak of summer and Scott desperately unlucky to arrive just after it. Temperatures depend on altitude as well as latitude and the Russian base Vostok at 78°28′S and 3400 m above sea level holds the world record for low temperature, −89.5 °C (Fig. 1.6). Wind speeds are generally low and precipitation extremely low. Direct measurement of snowfall is imprecise at best and the prevalence of drift on the continent makes it difficult or impossible. Between 3 and 7 cm rain equivalents seems to be likely – less than in most tropical deserts – but it is the extremely low moisture content of the air which makes the Antarctic plateau so highly desiccating.

As the slope of the ice cap steepens towards the coast, a different type of climate takes over, with strong and persistent katabatic winds averaging around 11 $m\,s^{-1}$ (39 $km\,h^{-1}$) and occasionally reaching 300 $km\,h^{-1}$ (Fig. 1.7). The coast itself has milder temperatures, dropping to around −20 °C in winter and rising to near zero in summer (Fig. 1.6). Over the sea the cold air from katabatic winds rises and is dissipated in turbulence, leaving conditions at the surface more tranquil. Cyclonic activity sometimes penetrates landwards bringing strong winds and precipitation. The weather is much

Fig. 1.7 Leaning on the wind, Adélie Land, Antarctica. (Photo F. Hurley, courtesy of the Scott Polar Research Institute.)

dependent on topography, which affects the incidence of katabatic winds and the amount of sea ice insulating the coast from the relatively warm sea. The US station, McMurdo, at 78°S on the Ross Sea has persistent sea ice but sunny summers with little snow so that the rocks are exposed for three or four months each year. The effect of topography is particularly marked on the Peninsula. On its west coast the Ukrainian station, Vernadsky (previously the British station, Faraday, at 65°15′S 64°15′W) has a mild maritime climate. At the same latitude on its east coast frigid conditions are maintained year round by cold water brought from higher latitudes by the Weddell gyre and an ice shelf extends out from the shore. Temperatures are some 4–6 °C colder than on the west coast.

Topography also produces 'oases', or dry valleys, which are a special feature of Antarctica. These are ice- and snow-free areas found at various points around the continent. The dry valleys of Victoria Land, accessible from McMurdo Station, have been intensively investigated as have those of the Bunger Hills in the vicinity of Mirny. Dry valleys exist where loss of snow and ice by *ablation* (i.e. removal by sublimation or run-off of melt water) exceeds addition by precipitation and movement of ice into the area. The configuration of the land surface must be such as to divert the flow of ice elsewhere and also to provide a precipitation shadow. The effect on the radiation balance of the resulting lowering of albedo has been mentioned (p. 5) and temperature fluctuations from around −38 °C to as much as +15.6 °C have been recorded. Winds are generally light but strong katabatic winds blow occasionally and wind-eroded rocks, *ventifacts*, are a striking feature of the dry valleys. The bare area tends to extend along the direction of the prevailing wind since debris is carried downwind and, being deposited on snow, decreases its albedo, promoting melting and exposure of bedrock. Extreme desiccation is a major factor for living organisms – the mummified remains of the occasional seals and penguins which stray into these cold deserts may remain for centuries. The annual precipitation is around 4.5 cm rain equivalents.

The maritime Antarctic, taken as the zone from 70°S northwards to 55°S including the Antarctic Peninsula and its associated islands together with adjacent archipelagos, falls in the domain of cyclones. Vernadsky Station has mean temperatures of around −10 °C in the winter rising to near zero in summer. There is more cloud than on the coasts of the main continent and winds are stronger. Sea ice usually disperses in early spring and re-forms in autumn. Signy Island (60°43′S 45°36′W), in the South Orkneys (Fig. 1.6, Orcadas), has much the same sea ice conditions and temperature range but has cloud for 80% of the summer and 60–80% of the winter. Annual precipitation, mostly snow but sometimes rain, amounts to 40 cm rain equivalents.

All these islands are well south of the Polar Front. Those in the vicinity of the Front are termed *sub-Antarctic*. South Georgia (*c.* 54°30′S 37°00′W) is some 350 km south of the Front. It has sub-zero temperatures every month of the year (Fig. 1.6) and permanent ice fields although temperatures of 15 °C are not uncommon. It receives an almost continuous series of atmospheric depressions and its climate may be summed up as generally cold, wet and cloudy, with strong winds, subject to abrupt change but without great seasonal variation. Katabatic winds, caused by cold air spilling down valleys, give rise to sudden squalls and whirlwinds, known as 'williwaws', which are frequent in some of the harbours. The southwest coast, exposed to the prevailing westerly winds, has a more rigorous climate than the sheltered northeast. These winds, heavily laden with water vapour, are forced to rise over the steep 3000 m spine of the island and in doing so expand, cool, and deposit their moisture as rain or snow. The then relatively dry air descending on the leeward side of the mountains is compressed and so warms up, producing rises in temperature of as much as 10 °C. These *Föhn* winds are an outstanding feature of South Georgia. Icebergs are common about its coasts but pack ice reaches it rarely. The French island of Kerguelen (*c.* 49°S 70°E) is about as far south as Paris is north, but its situation just on the Polar Front ensures a sub-Antarctic climate similar to that of South Georgia. Temperatures are again rather uniform, only falling a little below freezing in winter and rising little above 10 °C in summer but at 1050 m it has an ice cap giving rise to numerous glaciers. Like South Georgia it has strong westerly winds, fogs, rain, and snow. Icebergs are occasionally seen but its coastal waters are always free of ice.

1.3.4 Ultraviolet radiation

This is a meteorological element of considerable biological importance which has only recently received much attention (Karentz 1991). The solar radiation which impinges on the stratosphere has about 1% of its energy contained in the wavelengths 400 nm down to 200 nm (i.e. the ultraviolet, UV). This can be damaging to life since DNA absorbs maximally at 260 nm and wavelengths between 280 and 320 nm are capable of disrupting its structure and producing mutations. Other cell components, such as proteins and photosynthetic pigments, are also damaged by the shorter wavelengths of UV.

The atmosphere is efficient in filtering out wavelengths below 286 nm, largely because of the presence of ozone, which absorbs between 320 and 225 nm (UV-B down to 280 nm, UV-C below this), while allowing the less harmful longer wavelengths to penetrate. If the ozone in the atmosphere were concentrated into a discrete layer, this would have a thickness of 2.4–2.6 mm at the equator and 3.1–4.3 mm at 70′N, both at standard temperature and pressure. This seems, and actually is, a flimsy shield against a serious danger for living organisms. The concentration of ozone in the stratosphere, where it is most abundant, depends on a balance between its production, by the action of UV of wavelengths below 190 nm on oxygen, and its destruction, also by photochemical processes involving UV. It may be noted that these processes themselves utilize the energy of UV and result in warming of the stratosphere. Different balances between production and destruction explain the variation in ozone concentration with latitude.

In the 1970s it was realized that because chlorine has a catalytic effect on the destructive process, the release into the atmosphere of chlorofluorocarbons (Freons or CFCs), much used in industry as aerosol propellants and refrigerants, might affect ozone levels following transport of these substances into the stratosphere and their photolysis to release active chlorine. This possibility was taken seriously (Fogg 1992). The use of CFCs was restricted and more extensive and accurate monitoring of atmospheric ozone was put in train. This showed that between 1979 and 1986 there was a fall in the global mean total ozone of about 5% – an appreciable but not catastrophic decline. However, British Antarctic Survey scientists reported a dramatic decrease in total ozone over Halley Station, lasting over some six weeks in the austral spring of 1984. Values were down by about a third from those in 1957–77. At other times of the year the amounts were normal. This local 'hole' in the ozone layer had been missed by a sophisticated satellite-based world-wide monitoring system because this was programmed to discard low values which might be due to instrumental error. The existence of the 'hole' was confirmed on re-examination of the satellite data. Since then there has been a continued increase in the size of the 'hole' and the extent of the thinning (Jones and Shanklin 1995), which in 1995 amounted to about 30% and persisted into the summer. A similar, but smaller and less consistent, 'hole', has been found in the Arctic.

Such rapid, massive, and localized depletion of ozone was completely unexpected. The explanation lies in the pattern of circulation in the stratosphere in the polar regions during winter (Pyle *et al.* in Drewry *et al.* 1992; McIntyre in Wadhams *et al.* 1995). In both regions, strong westerly winds set up low pressure vortices in the stratosphere. The Antarctic vortex, centred over the pole, is the more stable whereas that in the Arctic is variable and its centre usually does not lie over the pole. The strong wind gradients around the vortices act as barriers to horizontal mixing and cut off masses of cold air from replenishment with ozone from outside. Since during the winter there is no UV falling on the polar regions to stimulate local production of ozone, processes eventually resulting in its destruction predominate. The declines in total ozone over Halley and the southern hemisphere are correlated with increases in CFC concentrations but low temperatures play an important part by providing ice crystals, on the surfaces of which the ozone-destroying reactions take place more readily. These ice crystals are visible as the nacreous clouds characteristic of the polar stratosphere. The increases in intensity at ground level of UV-B, to be expected from the falls in ozone concentration in the stratosphere, have been observed. When dynamical breakdown of the vortices occurs in the spring, the barrier to transport disappears, ozone levels are restored and the UV intensity falls to normal levels.

These findings have accentuated public anxiety about the global environment and have led to some international action to reduce industrial pollution of the atmosphere. The biological impacts of UV radiation are discussed later (p. 46) but a few further points may be made at this stage. One is that cloud reduces the intensity and impact of damaging UV radiation but it has no effect on the spectral composition of that radiation. Snow provides good protection to organisms covered by it by reflecting radiation and rapidly attenuating that which does penetrate. Clear ice and moderate depths of water do not afford protection to the same extent.

1.4 Matters of scale

1.4.1 Micro- and nanoclimates in terrestrial environments

Climates operate over wide areas but what is crucial for an organism is the combination of physical factors in its immediate vicinity (Walton in Laws 1984). These can differ enormously from those in the general environment. Warm sheltered spots and exposed prominences can be within short distances of each other. Slope is of obvious importance for the receipt of direct radiation but is of little significance in overcast conditions. Habitats in soil, among rocks or vegetation, provide irradiances, temperatures, humidities, and wind speeds quite different from those outside. Within short spaces of time conditions in a microhabitat may shift with shafts of sunlight or as thawing or freezing occur in its vicinity. Larger animals and plants are to varying extents able to select or modify their microenvironments but small organisms survive, or perish, according to the conditions in their immediate surroundings. How we define microclimate depends on the size of the organisms under consideration.

Measurements made by standard meteorological procedures give little, if any, information about these conditions. An example is provided by the two indigenous flowering plants of the Antarctic (p. 83) which are found in the maritime climatic zone but in very limited areas, within an otherwise ice-bound landscape, on sheltered north-facing slopes where they can get sufficient warmth and liquid water to be able to complete their reproductive cycles in a short summer. Correlation of their distribution with macroclimatic data is poor. Only measurements made in the air immediately surrounding the plants, among their leaves and in the soil underneath can give useful information about the conditions they need (see Fig. 2.2).

The instruments used for weather reporting are mostly unsuitable for micrometeorology. This is no place discuss the technicalities of instrumentation but it must be emphasized that the sophisticated equipment now available has to be used with understanding of its capabilities and limitations and with a clear idea of the sort of information that is required. This has not always happened. Wildly misleading results have been reported, even with such apparently straightforward measurements as that of temperature. Walton (1982 and in Laws 1984), in a critical discussion, has listed questions that must be pondered before any attempt is made to collect data. First, it must be decided which particular factors (e.g. temperature, humidity, etc.) are to be measured, and in what form (e.g. integrated daily total, maximum and minimum values, time beyond a particular threshold, etc.). It may be that the most useful thing is not a precisely definable physical measurement but a relatively crude determination which integrates the action of several variables, for example, data on wind-chill (p. 40) may be more informative than separate measurements of temperature and wind speed. Second, having taken these decisions, one has to choose the most suitable transducer. Physical conditions can usually be measured with greater precision than can biological responses and the degree of accuracy should be decided with this in mind. Then thought must be given to the frequency with which inputs are to be scanned. An upper limit will be set by the characteristics of the recorder but information should not be collected at the maximum rate on the assumption that redundant data can be discarded later. This may lead to more frequent failure of field equipment and complications in analysis of data. The most rapidly changing variable is usually radiation but by using an integrator it is possible to sample the input at a rate which is the same as that needed for more slowly changing factors such as soil temperature. Short-term observations give inadequate impressions of microclimates and continuous measurements over at least a year are desirable.

Operation of sophisticated equipment in polar regions is fraught with problems. Even if watertight containers are available, it is best to give batteries and recording equipment, not to mention those who service them, the extra protection of a tent or hut. Drifting snow has an exasperating ability to penetrate through the smallest gaps to cause short circuits. Low humidity allows static electricity to build up on masts and sensors to cause serious damage to circuits. Low temperature may cause plastic insulation to become brittle and crack, again leading to short circuits on subsequent thawing. Magnetic tape loggers are usually preferable but have the disadvantage that the tape may become brittle below −15 °C. Chart recorders must not depend on ink, which can freeze. The characteristics of electronic equipment change with temperature, and temperature coefficients should be determined so that allowance can be made for this. Under severe conditions it may be necessary to heat apparatus to keep the data recorded within a known range of error. Power supplies are crucial. Batteries must be chosen to give an ample safety margin. Solar panels, thermogenerators, or wind generators may be used to maintain batteries in charged condition, but bring their own problems. Biological hazards may be added to all this. Large mammals may wreak destruction. Burrows of small mammals and birds may displace buried sensors, rodents may gnaw unarmoured cable and birds may roost on solarimeters or destroy radiometer domes. Antarctic skuas can be a particular nuisance, pulling up cables, attacking instruments, and flying off with small objects which are left unattended.

What can be achieved is shown by data collected by a satellite-mediated station set up on Linnaeus Terrace (77°36′S 161°05′E) in the Asgard Range, southern Victoria Land, to monitor the environmental conditions affecting microbial communities in a dry valley (Friedmann *et al.* 1987). This is better described as nanoclimatology than microclimatology because it was necessary to determine gradients in some factors, such as temperature, well below the microclimate scale, over a few millimetres. The likelihood that biological activity was confined to brief periods spread over many months, together with the inaccessibility of the site, made it desirable that it should be a year-round unmanned station. The station set up operated for over three years, measuring rock temperatures, air temperatures, light, snow, wind, and moisture. The data were transmitted via the TIROS/NOAA series polar-orbiting satellites to the ARGOS centre in France. The biological significance of the results obtained are discussed later (p. 77).

1.4.2 Micro- and mesoscale habitats in aquatic environments

Small-scale variation in factors affecting organism are also important in aquatic habitats. Although distribution of the larger organisms – macroalgae, benthic invertebrates, fish, seals, whales, and seabirds – is mainly dependent on the conditions prevalent in large water masses, activities of microorganisms are affected by variations on scales from a few kilometres down to fractions of a millimetre.

In littoral and benthic habitats, microenvironments are determined largely, as on land, by the nature and configuration of the solid substratum. On a rocky shore, microhabitats may provide shelter from wave action, desiccation, insolation, and predators. Benthic sediments may develop particular chemical features, providing, for example, gradients of reducing activity if they become anaerobic. Open water is rarely homogeneous, being usually stratified vertically with respect to temperature, salinity, and concentrations of biologically important substances. Discontinuities occur in the horizontal plane at fronts. Microhabitats in this environment may be seasonal or ephemeral, others, those produced by turbulence, need not be fixed in space.

1.4.2.1 Gradients in still waters

If a water body is not stirred it will tend to stratify, because the surface layer heats up and becomes less dense, or because of the introduction of water of different salinity, or for other reasons. The pycnocline marking the boundary between water masses of different densities in polar waters has already been mentioned. Melting ice produces a surface layer of less saline water above a pycnocline of only a metre or so thickness, a situation which is particularly important for phytoplankton production (p. 191). Stratification may be well developed in sheltered lakes and especially in those protected from wind by ice cover. The water column inevitably has a gradient of irradiance intensity and quality from top to bottom. As a result of biological activity, sedimentation, and upward transport from bottom deposits, gradients may develop in concentrations of oxygen, nutrients, or other substances. If the water column is stabilized, different species come to dominate at different depths, each proliferating where it finds the conditions which suit it best. This can sometimes be seen by divers as water layers, perhaps only a centimetre in thickness, picked out by a dense growth of some pigmented microorganism. Interstitial water in a sediment or ice is more protected from mixing and here, chemical gradients may develop in which biologically significant changes may occur in a matter of millimetres. Clearly, to understand the ecology of such situations one needs to be able to make measurement of physical and chemical factors within small distances. Paralleling the instrumentation for micrometeorology there is a variety of sensors and ancillary equipment available for measurement of irradiance, temperature, salinity, and oxygen concentration in aquatic habitats. In sediments, it is usually most convenient to remove a core, with a minimum of disturbance, and make measurements in the laboratory.

1.4.2.2 Fronts in the sea

Fronts are narrow zones of demarcation appearing at the surface between different water masses (Fogg and Thake 1987). The Antarctic Polar Front is the grandest example, but smaller ones are frequent, especially in shallow coastal waters where tidal stirring may differentiate a water mass from adjacent stratified waters. Fronts between melt water from glaciers, which appears milky or pale blue because of the content of rock flour, and seawater, a clear green or indigo, are particularly striking. While the Polar Front is a permanent feature, these others may have only a seasonal existence, although they appear regularly in the same positions each year. Fronts have steep gradients in properties such as temperature, salinity, and nutrient concentrations across them and are of considerable biological significance. In studying fronts, satellite images or air photography give the best idea of their form but *in situ* measurements over short distances are needed to give the physical background necessary for biologists. Packages of sensors which give continuous readings as they are lowered to depth or towed horizontally to obtain transects are most useful here.

1.4.2.3 The scales of turbulence

Turbulence affects all aquatic organisms. For plankton its effects are immediate; although irradiance and nutrients are the actual determinants of phytoplankton growth, it is water movement which positions the cells in the intensity or concentration gradients and enables access to these essentials (Reynolds 1992). For larger organisms turbulence usually operates indirectly, but nonetheless is important, even for whales and seabirds, in determining the distribution of their food. In polar waters, which range from the roughest seas in the world, stirred by great ocean currents, to the quietest ice-bound lakes, turbulence is of particular interest.

Water movement is a continuum, varying in space and time, from circulations filling ocean basins, through gyres and eddies, down to molecular motion in which gradients of concentration occupy fractions of a millimetre and dissipate in milliseconds. The motive power for all these movements comes ultimately from the global changes which have been discussed in earlier sections (1.2.5, 1.2.6). Depending on wind speed, current shear, and stratification, timescales for cycling of plankton by turbulent eddies and mixing in the sea have been found to vary from about 0.5 to 100 hours for vertical displacements of 10 m. The mixing response in the surface layer to the onset of strong winds is rapid, taking a few hours during which cycling times are reduced to less than an hour. Large eddies transfer their energy to smaller ones and those around 100 km in diameter may be characterized by the time taken to transfer half their kinetic energy to eddies of half the original size. Comparing this time with the doubling time of the plankton one comes to the conclusion that such eddies can maintain within themselves populations of the more rapidly multiplying plankton species. Rings of 200–300 km can retain their characteristic phytoplankton flora for two to three years although the larger zooplankton, with long response times and migratory habits, tend to disperse. At around a critical diameter of 1 km, patches of phytoplankton disperse. Eddies thus play an important part in determining patchiness of phytoplankton in the sea and their internal circulation probably plays a part in holding it near the surface. Unfortunately for biological oceanographers, direct measurement of turbulence is extremely difficult and only approximate estimates of its intensity can be obtained from time-averaged fluctuations in the horizontal and vertical velocities observed at a particular point. Turbulence remains one of the least understood areas of classical physics and the variety of complicated non-linear behaviour is rich in situations which lead to chaos.

At the lower end of the scale turbulent energy is finally dissipated in molecular motion, i.e. thermal energy. This is a crucial frontier for microorganisms (Fogg 1991a). The transition of turbulent flow into the realm in which molecular motion predominates and flow become laminar is dependent on shear rate and viscosity. The ratio between these two is expressed in the dimensionless Reynolds number (R_e). The critical value for R_e in water lies between 500 and 2000, flow being laminar on the lower side, turbulent on the upper. The domain of low Reynolds numbers is one in which molecular diffusion is the dominant agency for transport of materials and in which inertia is irrelevant. Molecular diffusion over distances of around 100 μm, such as we are dealing with in this domain, is rapid and the circumstances that microorganisms have high surface/volume ratios and that their small radii of curvature have the effect of steepening concentration gradients, allows rapid exchange of materials between cell and water. For a cell 1 μm in diameter – about the size of an average bacterium – the flux of phosphate from a low concentration such as is usually found in natural waters is several hundred times what is needed to maintain maximum growth rate. Small organisms have a tremendous advantage over larger ones in competing for dissolved nutrients. Because the inertia of the cell becomes unimportant, its motility is of a different kind and serves different functions to those for larger forms. There is, in fact, a distinct break in form and function at a size of around 8 μm between smaller organisms, such as bacteria, picoplankton, and microflagellates, and the larger plankton diatoms and crustacea. Life in the domain of low Reynolds numbers is different from and to some extent separate from that in the world of turbulent flow but is nevertheless of basic importance in aquatic ecology. In the interstices of sea ice it becomes the predominant mode of existence. Unfortunately, there is a dearth of relevant rheological and shear rate data from natural waters which makes quantitative study uncertain. It would seem that since the viscosity of water increases at low temperatures, being at 0 °C twice what it is at 25 °C, that the extent of domains of laminar flow should be greater in polar as compared with tropical waters, but there is no information as to whether this is of significance.

1.5 Magnetic and electrical phenomena

Investigation of the earth's magnetic field in the vicinity of the poles was the principal attraction of the Arctic and Antarctic for early scientific expeditions. At the beginning of this century it began to be realized that electrical phenomena in the atmosphere have some relationship with the magnetic field and, in recent years, polar studies have made major contributions to the concept of geospace. Cusps in the magnetosphere, one over each pole, allow the charged particles of the solar wind to penetrate deeply into the polar atmospheres. One manifestation of this is in the aurora (Walton 1987; Pielou 1994). In the present state of our knowledge it seems that this uniquely polar situation is not of great significance for the life of these regions. Possibly, the fixation of molecular nitrogen produced by the greater electrical activity may support marginal increases in biological productivity. However, it is well established experimentally that some organisms react to magnetic fields. There is some evidence that variations in magnetic fields affect the time-course of plant growth. Birds and mammals may orientate themselves with reference to the earth's magnetic field so that migrations in the vicinity of the poles may have interesting features. There is a suggestion that magnetic disturbances around the poles may interact with brain activities and interfere with sleep in humans. It may therefore be premature to dismiss the special geophysical characteristics of the polar regions as of no interest to the biologist.

1.6 Changes during geological time: the ice ages

It is incorrect to thick of the polar regions as having always been as they are now. In both the Arctic and Antarctic, sedimentary rocks, still in more or less the same geographical position as when they were laid down, contain fossils of plants and animals characteristic of warmer climes and unlikely to have existed under conditions similar to the present ones (Cocks 1981). Existing habitats may not contain all the types of organism that one might expect and some give the impression of supporting immature communities which are still developing towards their climaxes as plants and animals migrate into them. It is necessary to look back through geological time if we are properly to understand the present ecology of polar habitats.

The scene was set by the drifting of continents according to the theory, now generally accepted, put forward by Wegener in 1912 (Cocks 1981; Walton 1987). Some 250 million years ago the Atlantic and Indian Oceans did not exist and a single land mass, Pangaea, was surrounded by the precursor of the present-day Pacific Ocean. Pangaea broke in two about 220 million years ago, the northern part fragmenting, with the pieces separating so that what are now the Siberian coast, Greenland, and the North American archipelago enclosed a sea area containing the North Pole. The northern edges of the continental plates on which these lands are carried form the physiographic borderlines of the Arctic Ocean. The movement of land masses into the vicinity of the North Pole had a profound effect on climate. The heat storage effect of the sea was reduced and winter cooling of the land brought about a fall in mean annual temperature leading to extensive glaciation by the beginning of the Pleistocene, 3 million years BP (Before Present).

The situation of the Arctic Ocean in a polar basin communicating with the world ocean by narrow channels has given rise to remarkable climatic fluctuations with a periodicity of about 100 000 years. This is explained simplistically by a theory put forward by Ewing and Donn (1956). Ice-free, the Arctic Ocean is the site of a permanent cyclone which moves moist air northwards, resulting in heavy precipitation and the build-up of ice caps on circumpolar land masses. This transfer of water from sea to land causes a fall in sea level, throttling the channels through which warm Atlantic water reaches the polar basin. Freezing of the Arctic Ocean follows and on the extended cold polar mass an anticyclone develops, driving dry easterly winds. The ice cover then begins to ablate and the release of water brings about a rise in sea level, allowing warm water into the polar basin with progressive melting of ice and eventually an open Arctic Ocean again. *Da capo.*

The southern part of Pangaea, known as Gondwana, drifted south and between 180 and 200 million years ago itself split up to give the pieces which now form South America, South Africa, India, Australia, and Antarctica. This left Antarctica in about the position which it still occupies, in the vicinity of the South Pole. The presence of substantial coal seams of Permian age, some 248 million years, in the Transantarctic Mountains is evidence that the new continent originally had a temperate climate. However, the gap between Antarctica and its neighbouring land masses widened as these drifted north, until around 25 million years ago it had opened sufficiently for the Circumpolar Current to become established and isolate the southern continent. This had a profound effect on the Antarctic climate (p. 12) and from this time its ice cap began to expand. Nevertheless, as shown by discoveries of fossil invertebrates and wood as far south as the Beardmore Glacier at 85°S, Antarctica was warmer at various times in the Pliocene, between 2 and 5 million years ago, than at present.

The succession of glaciations and interglacial periods which is still with us began about 3.2 million years ago. Before this, high alpine regions had most of the world's ice. The fluctuations of the ice have been mapped and dated from a variety of different kinds of evidence. Periglacial features such as moraines, eskers, drumlins, ice-scratched rocks, and drainage patterns, provide basic information. Cores from ice and from sediments from lakes and the seabed yield data which is often surprisingly detailed and precise; laminations in cores allow years to be counted; oxygen isotope ratios in 'fossil' water or in remains of foraminiferan shells give measures of temperatures at the time of deposition; diatoms, pollen grains and other microfossils give a picture of flora and fauna and help in dating (Barrett 1991; Lorius 1991; Thiede in Wadhams *et al.* 1995). Rock fragments from icebergs mark the occurrence and extent of periods of glaciation. Accompanying these glaciations there have been falls in sea level as water became locked up in ice, with corresponding rises in interglacials.

The cause of the onset of these glaciations is obscure. During the Pleistocene the glacial/interglacial fluctuations have shown a relation with the Milanković cycle – the cycle of variations, with a period of 120 000–150 000 years, in incident solar radiation resulting from perturbations in the earth's motion. Of the five or six major glaciations which the earth has experienced the last one is abnormal in affecting both poles and, at least latterly, the succession of glacials and interglacials have been synchronous in the Arctic and Antarctic. Greenland ice cores show 22 interstadial (warm) events in the time-span 105 000–20 000 BP. A core from the Antarctic shows nine interstadials in this period, these occurring whenever those in Greenland lasted more than 2000 years. This linkage between events at the two poles seems to have been mediated by partial deglaciation causing changes in sea level and ocean circulation.

2 The biological background

2.1 Introduction

This chapter deals with the effects of polar conditions on living systems in general, the ways in which microorganisms, plants, and animals are able to adapt to the stresses imposed, and how they come to be in these habitats.

Stress is a difficult concept to define in precise physiological terms. Radical change in photoperiod is one stress but cold – both in itself and through its secondary effects – is the factor which we think of first as inflicting greatest stress in polar habitats. The impact is rather different on small organisms, termed *ectotherms*, whose internal temperature conforms to that of their environment, and on larger plants, which to a limited extent can achieve higher temperatures than their surroundings, and warm-blooded animals. Among ectotherms the term *psychrophile* denotes a microorganism which is able to grow at or below 0 °C, with an optimum temperature at less than 15 °C and an upper limit at 20 °C. *Psychrotrophs* can also grow at around zero but show optimum growth at temperatures greater than 15 °C and have upper limits at about 40 °C. The distinction between psychrotrophs and *mesotrophs* is unsatisfactory since microorganisms classified as mesotrophs may be persuaded, given suitable conditions, to grow below their supposed lower limit of 5 °C. The adaptations to cold shown by higher plants and animals, as well as by microorganisms, may be *genotypic*, inherited. *Phenotypic* adaptation, resulting from interaction between the genetic constitution of the individual and its environment, is termed *acclimatization* if it is a response in the field to a whole complex of cold-related factors. *Acclimation* refers to laboratory-induced adaptation to a single variable.

Investigation of how organisms react to and withstand polar stresses is of more than academic interest, being relevant to the breeding of crop plants and domestic animals and their management in cold climates. Studies of the effects of cold on microorganisms are valuable in controlling spoilage of food during refrigeration, in the extended storage of viable biological material at low temperatures, and in identifying strains of microorganisms which might be exploited in biotechnology to avoid the expense of providing elevated temperatures.

2.2 Biological rhythms in the polar environment

2.2.1 Photoperiodism

Rhythms of activity, development, behaviour, and reproduction are usually geared to periodicities in the environment, particularly to those in light and dark, heat and cold, and tides. Some degree of synchrony between these biological and physical cycles must occur if organisms are to succeed. Timing, as, for example, in producing progeny when the physical conditions make food available, is a crucial factor in the life cycle strategies of polar animals. Many species have timing mechanisms to spread development over several years, rather than risk concentrating it all in one short favourable period.

An overriding periodicity, since it limits the time during which the primary production by photosynthesis upon which everything else depends takes place, is the cycle of light and dark. The mechanism by which organisms keep track of the progress of the seasons by response to duration of day and night was first studied in plants. These fall broadly into three classes according to their requirements for flower initiation, namely short-day, long-day, and day-neutral plants. If they are to flower, short-day plants must have an uninterrupted dark period of at least a certain length, usually about eight hours in the twenty-four. Most spring- and autumn-flowering plants of temperate regions are of this kind. Long-day plants, which include all summer-flowering plants of temperate regions, do not need a dark period but require twelve hours or more of light in the daily cycle. Little work has been done on the photoperiodicity of polar plants but it seems to be of the long-day type. Vegetative growth and dormancy also show photoperiodic effects and there are indications that short days induce cold-hardening. Parallel responses to those in flowering plants are found in ferns, mosses, and seaweeds. The inability of some mosses to maintain photosynthesis under continuous light may account for their absence in high latitudes (Kallio and Valanne 1975).

Photoperiodic effects are found in some Arctic insects, diapause (suspension of development) being brought about by short days. They are also shown in the migrations, reproduction, and changes in coat or plumage of mammals and birds. The change to white in winter in the Arctic hare (*Lepus arcticus*)* and in ptarmigans (*Lagopus mutus* and *L. lagopus*) is photoperiodic and can be induced out of season by exposure to short days. In reindeer (*Rangifer tarandus*, also known as caribou), levels of growth hormone and testosterone show seasonal variations which seem ultimately related to external factors such as photoperiod. Northern hemisphere reindeer introduced into the Antarctic (p. 224) adapted well to a reversal of the photoperiodic cycle.

*Common names will be used in this book where possible but the scientific name will be given on first mention.

Photoperiodism is often complicated by temperature effects. *Colobanthus subulatus*, a pearlwort native in South Georgia, is a long-day plant but needs cold pretreatment for flower initiation. It seems as if the very early inflorescences which it produces in September must have been initiated in the previous season with the cold pretreatment happening in the winter before that. In *homoiothermic* – warm-blooded – animals the complicating effects of ambient temperature are less than they are for plants and illumination, which acts through eye stimulation releasing hormones under pituitary control, has more clear-cut influence.

2.2.2 Circadian rhythms

Circadian rhythms in metabolism (i.e. variations in activity with a periodicity of around 24 hours), seem to be involved in the biochemical mechanisms which produce photoperiodic response, as a means of counting the passage of time. These rhythms, which are found at all levels of organization, are endogenous (i.e. continuing to a large extent independently of environmental changes), although the clock may be stalled and reset by sudden alterations in conditions such as temperature.

In plants, photosynthesis commonly shows a circadian rhythm. Transferred from a light/dark regime to continuous light, plants may show for some days afterwards a minimum in photosynthesis at the time corresponding to that of the original dark period. However, it is difficult to establish that similar endogenous rhythms are important under field conditions. Measurement of carbon dioxide fixation by the marsh plant *Caltha palustris* on the Murmansk coast, around 69°N 35°E, during the continuous daylight of the Arctic summer showed periodicity with a maximum near noon and a minimum near midnight. However, the variations were correlated with radiation intensity and there was no evidence of an endogenous rhythm (Pisek 1960). Diurnal variations in phytoplankton photosynthesis in polar regions seem likewise related to irradiance rather than to a biological clock (p. 120).

Among animals, Arctic bees show diurnal rhythms. Birds might be expected to show activity throughout the 24 hours of continuous daylight but, in fact, usually retain some degree of diurnal rhythm (Sage 1986). The resting period of light active passerines – song birds – shifts further into the evening with lengthening light period put there is usually a resting period around midnight. An overriding effect of a circadian rhythm is evident in the behaviour of the snow buntings (*Plectrophenax nivalis*) in the narrow valley of Longyearbyen (78°12′N 15°40′E) in Svalbard. Because of its orientation between high mountains, this valley only receives direct sunlight around midnight and is otherwise in shadow. Nevertheless, the resident birds sleep at midnight as do those in open tundra. Different species can, however, show different timings of activity and inactivity. Non-passerines, such as ducks, auks, gulls, and terns, are capable of continuous activity although they show circadian rhythms in other respects.

Circadian rhythms occur in man and the polar regions are obvious places for study of the effects of unusual schedules or time zone shifts, without confinement of subjects to controlled environment rooms. Interest in this began in the Arctic as far back as 1910. It is a general experience that the photoperiods of high latitudes have an upsetting effect, resulting for some in serious impairment of sleep cycles and sleep structure, peaking at the times of continuous daylight or the middle of the polar night. Anomalous responses to medication may be induced. In an elaborate study at the South Pole sleep and activity patterns, haematological and cardiopulmonary data, and the acute effects of oxygen shortage (the Pole is at an altitude of 2912 m), were examined. A surprising finding was that stage 4 (slow-wave) sleep was almost completely lost by the end of the austral winter and failed to return six months after the subject was back in the United States. Nothing similar was found to happen at the British station at Halley, further north. The discrepancy may be related to differences in group behaviour, such as having fixed mealtimes – a British rather than an American habit – which may act as synchronizers to adapt circadian rhythms to a 24-hour day (Wortmann 1995).

The role of photoperiodism and endogenous rhythms in the lives of polar organisms needs to be looked at more fully. Presumably, they play a vital part in gearing life cycles to the stringencies of the polar seasons.

2.3 Life at low temperatures

2.3.1 Effects of low temperatures on cell physiology

As temperature falls the proportion of molecules which are in a sufficiently activated state to undergo reaction diminishes and the rate of chemical transformation is reduced. The Arrhenius equation which expresses this provides a fundamental approach to the study of the effect of temperature on an isolated chemical reaction and can be applied to enzyme-catalysed processes. However, it is inappropriately applied to complicated living systems and for these it is more convenient to use an empirical temperature coefficient, Q_{10} ($Q_{10} = (k_1/k_2)^{10/(t_1-t_2)}$), derived from the rates of a process, k_1 and k_2, measured at temperatures of t_1 and t_2 °C, respectively). This can be applied to any sort of activity. The effects of chilling on the processes going on in an organism are complicated by biochemical interactions but it is usually found that Q_{10}s are between 2 and 3, so that rates at 0 °C are only a half, or less, of what they are at 10 °C. So long as liquid water is available, which it may be down to −40 °C because of supercooling (or, more correctly, *undercooling*), biochemical processes should continue at lower, but still perceptible, rates at temperatures below zero.

Nevertheless, cold may kill and few tropical or temperate plants or animals can survive polar temperatures. Broadly speaking, this lethal effect arises because different processes have different temperature coefficients so that the balance between them becomes distorted as temperature falls. The efficiency with which enzyme reactions are coupled together may be reduced and adenosine triphosphate (ATP) production may become insufficient to meet the energy requirements of the cell. Enzymes themselves undergo changes and become inactivated. Denaturation of proteins at high temperatures is familiar to anyone who has boiled an egg but it may equally happen at low temperatures, although this has been much less studied. The particular folding pattern of its peptide chain, on which the catalytic activity of the enzyme protein depends, is determined by hydrophobic interactions, which decrease in strength, and hydrogen bonding, which increases in strength, on chilling. The relation between them becomes disturbed, the molecule unfolds and becomes inactive (Jaenicke in Laws and Franks 1990).

There has not been much research on the enzymes of cold-adapted organisms. One way of countering the effects of low temperature is to compensate for the reduced activity of a key enzyme by producing more of it. This happens in some Antarctic fish, especially with enzymes involved in aerobic respiration (Eastman 1993). High rates of photosynthesis in Arctic flowering plants are made possible by higher than usual concentrations of the rate-limiting photosynthetic enzyme, ribulose bisphosphate carboxylase (RuBP carboxylase/oxygenase). The leaves of Arctic plants have, as a consequence, higher nitrogen contents than those of temperate plants. Genotypic modifications in the molecular configurations of enzymes in polar organisms, giving them greater efficiency at low temperatures, have also been found (Russell in Laws and Franks 1990). An enzyme in a psychrophilic organism may have the same qualitative properties as its isozyme in a mesotroph but differ in quantitative characteristics. Thus, RuBP-ase isolated from Antarctic diatoms has the same activation energy as that from a species from a temperate habitat, but, surprisingly, the rate of catalysis by the psychrophile enzyme reaches a maximum at 50 °C compared with 40 °C with that from the mesotroph. However, the affinity of the psychrophilic enzyme for its substrate is greatest at 4.5 °C as compared with 20 °C for the mesotrophic form and it is this which is evidently of biological advantage in cold waters. The expectation that enzymes of ectotherms adapted to cold conditions should have lower activation energies than those from warmer climates has been verified for some higher plant enzymes and for protein-cleaving enzymes of planktonic and benthic crustaceans in Antarctic waters.

Another damaging effect of chill is derangement of the cell membrane, the integrity of which is essential for the control of exchange of substances between the cell and its surroundings. This membrane is a liquid lipid layer modified by the presence of protein. The principal lipid molecules are arranged at right angles to the plane of the membrane to form a bimolecular layer with the non-polar hydrophobic groups on the outside. Some 15–25% of the total lipid does not form bilayers but instead the molecules are grouped into aggregates – *micelles*. The protein consists of various enzymes which porter substances across the membrane and also play a part in stabilizing it. Few general principles relating to membrane stability have been agreed but it seems that a usual occurrence when temperature is lowered is a transition from a liquid crystal structure to that of a gel, with damage showing itself in leakage of solutes from the cell (Williams in Laws and Franks 1990). The lipid composition of whole cells of ectotherms changes, particularly with a shift from saturated to unsaturated fatty acids, when they are grown at near zero temperatures rather than in warmer conditions. This phenotypic change parallels genetic differences. Psychrophiles, as a group, have higher proportions of unsaturated fatty acids (52%) than do mesotrophs (37%) and thermophiles (10%). Psychrophiles also have more short chain fatty acids. Both these changes could lower the temperature at which the liquid crystal/gel phase transition of the membrane occurs and perhaps might contribute to its chill resistance. However, the weak interactions between the different membrane components are sensitive to a complex of factors, including pH, ionic concentrations, and hydration, as well as to temperature, and it is not possible at present to be sure that lipid composition is of major importance. One of the effects of cryoprotectants – substances such as glycerol and proline which protect against freezing (p. 36) – is to increase membrane stability and hence chill resistance. They seem to do this by acting as 'water replacement agents' which maintain a balance between membrane components similar to that which exists under normal physiological conditions.

It had long been supposed that polar *poikilotherms* – cold-blooded animals – adapt to cold by having elevated basic (*routine*) metabolic rates as compared with those of similar animals from warmer environments. This would diminish the amount of energy available for growth and reproduction and so conveniently account for the slow growth rates, delayed maturation, and prolonged gametogenesis characteristic of polar organisms. However, the data on which this theory was based seem to have been obtained by faulty experimental procedures. There is now much physiological and biochemical evidence to refute it (Clarke 1983). With Antarctic marine invertebrates, for example, under properly controlled experimental conditions there is no detectable elevation of the rate of routine metabolism above that to be expected at low temperatures. The slow rates of growth and reproduction are puzzling but presumably are due to an overall reduction in energy utilization in response to conditions in the environment (Johnston in Laws and Franks 1990).

2.3.2 Effects of freezing and freeze resistance

The effects of freezing are distinct from those of low temperature *per se*. Chill does not necessarily involve the separation of ice nor the changes in water-soluble components which inevitably accompany freezing. Ice is formed in the extracellular phase before freezing occurs inside the cell. This releases latent heat and helps to buffer the cytoplasm against fall in temperature and freezing. Freezing requires the presence of nuclei for ice crystal formation. These can form spontaneously in the course of the random movements of molecules in the liquid, but, except in the region of −40 °C, usually decay before crystal growth is initiated. In living systems nucleation is catalysed by particulate matter presenting particular molecular configurations. It is unfortunate for invertebrates that nucleators are present in food materials so that freezing conditions are best faced with an empty gut.

When ice forms in the immediate vicinity of unicellular organisms or in the intercellular fluids of multicellular organisms, solutes are excluded from the crystals and so their concentration in the remaining liquid is increased. The osmotic stress which results is the immediate and most injurious consequence of freezing. Water is drawn from the cell, the magnitude of the efflux depending on the excess solute concentrations in the extracellular phase and the permeability of the cell membrane. Under field conditions cooling rates are usually slow and high membrane permeability to solutes ensures that osmotic equilibrium is maintained as the extracellular fluid becomes gradually more concentrated. The reduction in cell volume as water is withdrawn may lead to injury by impairing the resilience of the membrane so that the cell bursts on rewarming. Under more extreme conditions cell dehydration and perhaps internal freezing may cause mechanical deformation and leakage of cytoplasmic solutes. Another possibility is that as substrates become more concentrated the rates of enzyme reactions are speeded up and a deleterious imbalance of metabolic pattern set up.

These injurious effects may be minimized in various ways. Cold-acclimatized microorganisms, plants, and animals, produce substances which counter the effects of freezing. Osmotic imbalance can be reduced by the production within the cell of 'compatible' solutes, a process known as osmoregulation. These substances, which include free amino acids and low molecular weight polyhydroxy compounds such as sugars, must, of course, be soluble and metabolically inactive. Additionally, they appear to have a role in stabilizing proteins. Production of compatible solutes may take a few minutes in microorganisms but freeze resistance takes several days to develop in insects and weeks in large plants. Changes in membrane elasticity may occur during cold acclimation, making cells more able to withstand contraction and expansion stresses. Plant protoplasts, for example, can develop tolerance to stretching as much as three times greater than that they possess normally. In certain yeasts and snow algae this correlates with the appearance, as seen by electron microscopy, of numerous folds in the membrane, which are not present in unadapted cells and which seem as though they should allow expansion or contraction to be taken up without stress. Effective water management is a basic requirement in all cold survival strategies (Franks *et al.* in Laws and Franks 1990).

2.3.3 Avoidance of chill and keeping warm

Complete avoidance or inhibition of nucleation at sub-zero temperatures prevents freezing. This is a thermodynamically unstable state – hickory trees are able to withstand cooling down to about −45 °C but then freezing happens suddenly and they are killed. Marine fish in polar regions have blood freezing points of between −0.9 and −1.0 °C but can live in deep water at a temperature of −1.8 °C because the absence of ice nuclei allows them to remain in the supercooled state. If ice is introduced into their vicinity they instantly freeze as nuclei diffuse from ice to fish. Antifreezes occur universally in Antarctic fish and less commonly in Arctic fish, some of which only produce them during the winter. They are not specific to particular taxonomic groups. Antarctic fish have had more time, about 25 million years (see p. 203) to evolve the necessary biochemical mechanisms, as against only two to three million years for those in the Arctic (Johnston in Laws and Franks 1990; Eastman 1993). Antifreeze substances are also present in some plants (p. 138) and some insects (p. 43). They include glycerol, polyhydric alcohols, and glycoproteins. The action of the low molecular weight substances, such as glycerol, is *colligative*, that is to say proportional to the molecular concentration of the solute. In terms of the fresh weight of the organism a colligative antifreeze may reach a concentration of 3–6%, or even as high as 14%. The high molecular weight substances have low osmotic activity and their action is non-colligative. They act by becoming adsorbed on ice crystals and inhibiting their growth. The long straight fronts in the growing crystal are interrupted by the adsorbed molecules and divided into smaller fronts. These become curved, causing increase in the free surface energy of the crystal and hence a lower temperature is required for freezing (Eastman 1993).

Form and function plays a part in ameliorating temperatures for plants. A cushion or turf not only provides an insulating layer protecting the lower parts of the plants (and, incidentally, a better environment for animals) but also acts as a trap for radiant energy. Absorbed radiant energy which is not used for photosynthesis or evaporation is re-emitted at longer wavelengths but, because of the structure of the plants, much is reabsorbed by them. In the sun, the temperature inside a cushion or turf may rise to 10 °C or more above that of the ambient air. Cloches have a similar effect (Figs 2.1, 2.2, 2.4). Phototropism may keep a plant organ orientated so as to receive the maximum amount of radiant energy. The glacier buttercup (*Ranunculus glacialis*) has flowers which swivel on their stems to face the sun as it crosses the sky. The flower itself acts as a parabolic reflector, directing the rays to its centre and, presumably, providing a more favourable temperature for the development of ovaries and stamens as well as attracting pollinating insects.

Insects are the largest metazoan animals to tolerate low body temperatures and to survive freezing. The larger marine invertebrates are not exposed to

Fig. 2.1 Experimental manipulation of high Arctic polar desert, with patchy vegetation dominated by *Dryas octapetala*, using plastic cloches to ameliorate temperature at Ny-Ålesund, Svalbard. (Courtesy of Professor T. V. Callaghan, photo Jens Busch.)

temperatures below −1.9 °C in their normal habitats. Arctic insects are mostly darker, smaller, and hairier than their temperate counterparts, these being features maximizing radiative warming. Their activity patterns take advantage of favourable microclimates and unfavourable periods are passed in dormant condition. Bumblebees, large insects with dense insulating 'fur', can raise their body temperature to as much as 35 °C by shivering of their flight muscles. This enables them to remain active in cool weather and to hasten development of eggs and larvae in the nest.

Mammals and birds avoid the problems faced by ectotherms by using heat released by metabolic processes, in conjunction with insulation and various physiological and behavioural mechanisms which conserve heat, to maintain body temperatures between 36 and 41 °C. Heat has often to be conserved against steep temperature gradients and requires a correspondingly ample supply of food or internal energy-yielding reserves. An extreme example is the male emperor penguin (*Aptenodytes forsteri*), which survives 105–115 days of winter, without feeding, at temperatures which may fall to −48 °C (p. 167). Aquatic mammals are in better situation because the sea does not fall below freezing point and, since it supports a greater bulk of body, allows low surface/volume ratios to reduce heat loss. Hibernation, by allowing body temperature to approach that of the surroundings, enables large savings in energy-yielding reserves but is not typically resorted to by polar animals. Perhaps this is simply because it is only possible in frost-free nooks and these are scarce. Breeding female polar bears (*Ursus maritimus*) enter a limited winter sleep in which body temperature falls for a few days

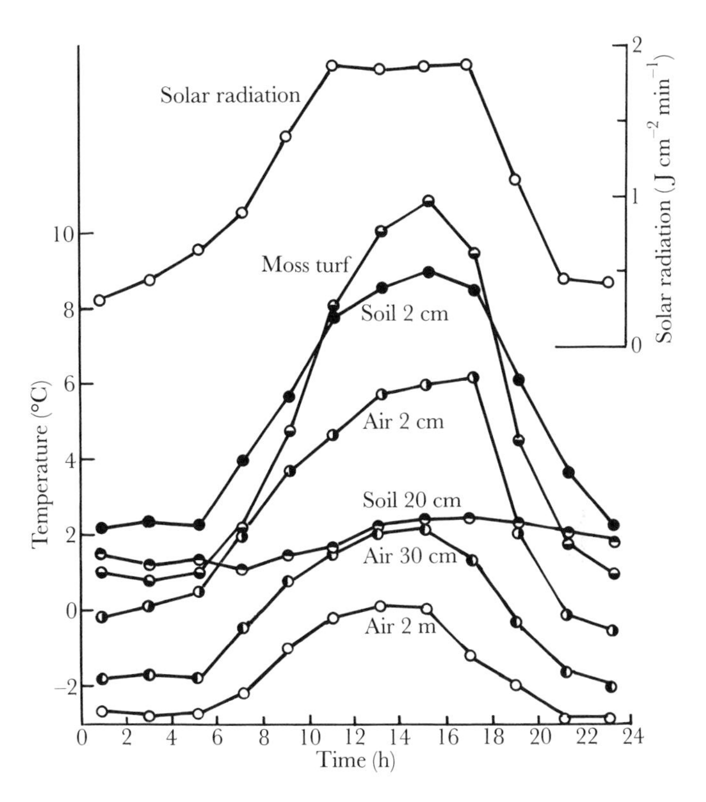

Fig. 2.2 Mean temperatures at 2-hour intervals over 5 days in, above, and below a moss turf (*Bryum argenteum*) in the frigid-Antarctic on Ross Island. (Redrawn after Longton 1988, by courtesy of the author and Cambridge University Press.)

at a time, but this is not true hibernation. Many birds and mammals avoid the hazard of a polar winter and find a more congenial habitat for breeding by migrating to warmer places. This, too, may involve considerable expenditure of energy. There is some tendency for polar birds and mammals to be larger than their temperate or tropical counterparts but other factors than heat conservation by reducing surface are involved (e.g. the need for completing the reproductive cycle as quickly as possible makes small size desirable), and there are many exceptions to this rule. Finally, it should not be overlooked that conservation of body temperature carries the danger of occasional overheating. Animals may sleek their fur or feathers, pant, employ circulatory bypasses to their extremities, or have bare or lightly insulated areas which they can expose, to cool themselves when necessary.

Thawing may not be the end of an organism's suffering, this too has its hazards. Cell membranes may be in an unstable condition and unequal to sudden change in osmotic stress. Freeze–thaw cycles are powerful in disintegrating rock and organisms cannot altogether escape mechanical disruption. In Signy Island soils there is a spring peak in microbial activity in response to organic substances released by thawing from damaged cells.

2.3.4 Wind-chill

Some effects of wind – mechanical damage, snow drift, abrasion by suspended ice crystals or rock particles, increased evaporation – perhaps do not require much discussion here, but the concept of wind-chill does. The physiological effects of cold depend on the rate of loss of heat from the organism. Loss by convection/conduction, as distinct from radiative loss, is determined by the temperature gradient between it and the ambient air. This gradient is steepened if the air is moving. For a given air temperature the loss will be greater as wind speed increases. An index of this effect is based on experiments on heat loss from dry uninsulated surfaces. This *wind-chill factor* is expressed either as equivalent temperature reduction (Fig. 2.3) or as additional heat loss per unit time. It is used extensively but has defects from the physiological point of view. If the surface in question is damp then more heat will be lost than the index shows, because of evaporation of water, but the temperature of the surface cannot fall below the wet-bulb temperature of the air – below this point heat is added by condensation of water. During storms, the effective cooling power of the wind is greatly increased by the ice and snow it carries. The index also takes no account of radiation or insulation. Animals with wind-resistant plumage or fur, and humans in wind-proofs, may be scarcely affected, even in strong winds. There is little correlation between freezing of uninsulated finger skin and wind-chill index because varying degrees of supercooling, cold-induced vasodilation, and skin moisture, complicate the situation. Nevertheless, the

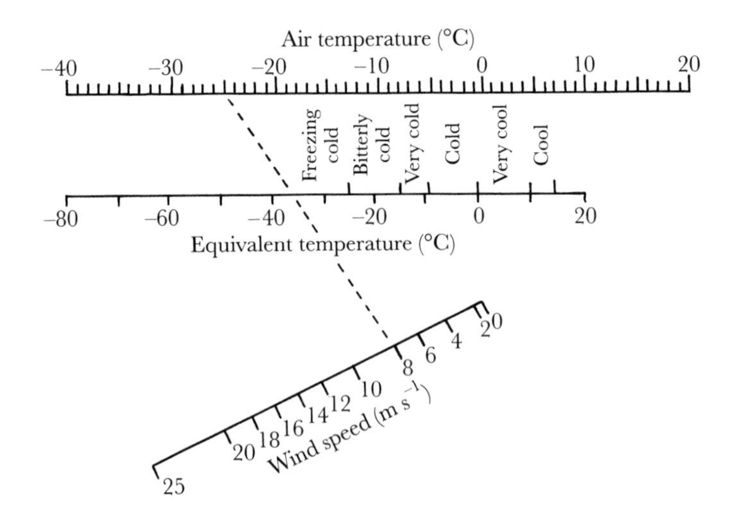

Fig. 2.3 A nomogram for determining wind-chill in terms of equivalent temperature. Equivalent temperature is read off by drawing a line between actual air temperature and wind speed, as illustrated by the dotted line in the figure which shows an air temperature of −25 °C and wind speed of 8 m s^{-1} giving an equivalent temperature of −36 °C (i.e. freezing cold with danger of frostbite to exposed flesh). (Redrawn from Rees 1993, by courtesy of the author and the editor of *Polar Record*.)

wind-chill index, used with discretion, remains a useful guide to polar conditions, especially as they affect bird mortality and the comfort of human beings. Among emperor penguins wind-chill in the period between leaving the brood pouch and learning to huddle (p. 168) seems to be the major cause of chick mortality, accounting for 80% of the loss.

An index of survival time outdoors in extreme cold, based on body–atmosphere energy budget modelling procedures, is more informative about effects on humans than the wind-chill factor. The index is the calculated time for a fall in core temperature, in an inactive healthy subject in full polar clothing, from 37 to 27 °C. The severest Antarctic conditions become life-threatening after 20 minutes. At most stations on the continent, a similar point would be reached in two hours of outdoor exposure in winter. Conditions at all coastal stations in summer are mild enough to allow core temperature to be maintained (de Freitas and Symon 1987).

2.4 Desiccation

For the maintenance of life, exchange of substances, including water, across the cell membrane is essential. If the water potential is less outside (e.g. if the concentration of osmotically active substances is greater than it is inside), then water is inevitably lost from a cell. The gradient in water potential between a cell grown in freshwater then transferred to seawater is equivalent to an osmotic stress of around 23 bar (1 bar = 10^5 Pascals ~ 1 standard atmosphere). Organisms exposed to the air may be subjected to much greater stresses. When the air is 90% saturated with water vapour at 20 °C the difference in water potential between it and a moist surface at the same temperature is 140 bar. When it is half saturated, as it may be in an English summer, it is as much as 924 bar. Relative humidities in the dry valleys of Antarctica are usually low, 30–40% in winter, rising to 80% in summer or even to 100% in the rare showers of light rain. Relative humidity falls to less than 10% when katabatic winds, already containing little moisture, warm up as they blow down-valley. This would result in a water potential deficit of about 800 bar at 0 °C.

Freshwater habitats present no problem but water remaining unfrozen in polar regions is often highly saline, with water potential deficits up to 200 bar, and therefore unavailable.

Lack of water is generally more limiting for terrestrial life than is low temperature.

2.4.1 Effects of desiccation on cell physiology

Resistance to desiccation has been little investigated at the cellular level. As might be expected from the crucial involvement of water relations in resistance to freezing, there is a general correlation of the abilities to withstand drought and low temperatures. This is seen in the tolerance to both these stresses, which is generally characteristic of the cyanobacteria and the chlorococcalean green algae, in contrast to the sensitivity of other groups of photosynthetic microorganisms. There is a genetic element in resistance to desiccation; Antarctic cyanobacteria and unicellular green algae appear to have a greater tolerance to freeze-drying than their temperate latitude counterparts. The conditions under which drying and rehydration take place have a great effect on whether cells will survive. The stage of growth, the rate of drying, and whether it takes place in light or dark, all seem important for one or other species but it is difficult to discern any general pattern. Metabolism is retarded in desiccated cells; cyanobacteria and lichens show no perceptible photosynthesis when dry, although they remain viable and may resume photosynthetic activity quickly on rewetting. The respiration of *Phormidium*, unlike that of another cyanobacterium, *Nostoc*, continues after photosynthesis is inhibited, which must deplete carbon reserves and possibly it is this which slows down recovery on rewetting of mats consisting predominantly of *Phormidium* (Hawes *et al.* 1992). A lipid pellicle is sometimes produced around microbial cells on drying and mucilaginous sheaths or thick cell walls, acting as reservoirs, will also reduce the rate of water loss from the cells themselves, giving time for physiological adaptation. Total lipid increases in water-stressed microbial cells but this does not seem to be related to desiccation resistance, although individual lipid components may be concerned in providing resistance.

2.4.2 Effects of desiccation on whole organisms

Dehydration, whether brought about by low relative humidity, freezing, or high salinity, is something the microflora of polar regions must withstand for long periods. The most severe test, no doubt, is during the colonization stage when cells are transported in the air, but viable microorganisms do arrive by this route (p. 50).

Plants are particularly affected by water deficiency since exposure of extensive areas of water-imbibed cell surface is required for carbon dioxide intake from the air for photosynthesis. With frozen roots, and shoots exposed to sun and wind, a plant quickly becomes water-stressed. Flowering plants in the sub-Arctic have morphological characteristics, such as reduced or fleshy leaves, usually associated with growth in water-deficient conditions (i.e. they are *xeromorphic*). Examples are some *Saxifragas* and crowberries (*Empetrum* spp.). A capacity to adapt metabolism to water stress is essential for the existence of most terrestrial plants. Provided that withdrawal of water is gradual, osmotic adjustment may be achieved via concentration of inorganic ions and organic solutes. Mosses and lichens, which are the predominating components of polar land floras are, as classes, characteristically desiccation-resistant and the polar forms remarkably so. Thallus morphology is important in reducing water loss and crustose or foliose habits in lichens and aggregation of mosses into mats or cushions (Fig. 2.4), decrease evaporation. Net photosynthesis occurs in polar lichens and mosses over a water content range of 100–400% of dry weight (Longton 1988). Some mosses and lichens remain viable after years at water contents of 5–10%. Tolerance to desiccation is related to habitat; many mosses growing in wet places are killed by drying for only a few hours. There is also seasonal variation.

Terrestrial invertebrates become inactivated if desiccated and their distribution is largely determined by water availability. The high degree of cold tolerance which polar forms have suggests a parallel resistance to desiccation. Indeed, the effects of desiccation and low temperature are closely intertwined. The springtail *Cryptopygus antarcticus* can survive a reduction in total body water from 60–40% of fresh weight, representing a fivefold increase in solute concentration assuming no osmoregulation. In the dehydrated state it has a remarkable ability to take in water but this is done at some sacrifice of cold tolerance. The Antarctic mite *Alaskozetes antarcticus* (Fig. 2.5) is very resistant to desiccation; on drying to a water content of 60% of fresh weight there is a significant accumulation of glycerol, which is an antifreeze (p. 37). A degree of mobility and a capacity for encystment or dormancy also help in evading drought conditions. Birds and mammals are able to search out liquid water and presumably do not have a desiccation-resistant type of cell physiology.

Desiccation in an arid environment may be avoided by occupying an enclosure within which water can be conserved. Continued life within such

Fig. 2.4 Moss cushion (*Dicranum* sp. with lichens, *Usnea* and *Stereocaulon* spp.), fell field, Signy Island. (Photo G. E. Fogg.)

Fig. 2.5 An Antarctic mite, *Alaskozetes antarcticus*. Although only about 0.2 mm in length it is one of the largest terrestrial invertebrates in Antarctica. (Courtesy of Dr W. Block and the British Antarctic Survey.)

a niche is only possible if there is an input of energy and cycling of materials, that is, there must be an association of photosynthetic activity – producing organic matter and oxygen and assimilating carbon dioxide – with heterotrophy – consuming organic matter and oxygen and producing carbon dioxide. This was achieved in the sealed containers containing living ferns which graced the dimly lit Victorian drawing room. Water, transpired by the foliage, condensed on the glass and ran back into the soil which sustained enough heterotrophic activity to balance the low net photosynthesis. Analogous microcosms, almost self-contained communities of algae and lichens with yeasts and bacteria, are found in the micropores of rocks in the dry valleys of Antarctica (p. 75).

2.5 The effects of radiation

2.5.1 Light

Light is, strictly speaking, radiation visible to the human eye but the word can be used without too much inaccuracy to denote radiation visible to other animals and also photosynthetically active radiation. The quality and range in intensity and the reactions of plants and animals to these are much the same in polar regions as elsewhere. However, the relations of photosynthetic activity to light intensity are of particular importance in many polar habitats and must be outlined. The rate of arrival of radiant energy per unit area of surface is properly termed *irradiance*, although *intensity* is generally used, loosely, to mean the same thing. Irradiance is measured in joules ($J m^{-2} s^{-1}$), watts ($W m^{-2}$), that is, in terms of energy, or photon flux density (moles quanta or einsteins $m^{-2} s^{-1}$), that is, in number of quanta. The energy of a quantum is dependent on its wavelength but for full sunlight a conversion factor of $1 J m^{-2} s^{-1} = 4.6 \mu mol$ quanta $m^{-2} s^{-1}$ is a reasonable approximation. Photon flux density will generally be used in this book and it will be helpful in visualizing light conditions to remember that the photosynthetically active component of full sunlight amounts to roughly $2000 \mu mol$ quanta $m^{-2} s^{-1}$.

The relation of rate of photosynthesis, usually measured as oxygen output or carbon dioxide uptake, to irradiance is shown in Fig. 2.6. In darkness, the uptake of oxygen and release of carbon dioxide in respiration are the dominant exchanges. With low levels of irradiance these exchanges diminish and at a particular level, the *compensation point,* become zero as photosynthesis and respiration cancel each other out. Beyond this point the rates of oxygen output and carbon dioxide uptake increase linearly with irradiance. Light is at first the limiting factor and variation in concentration of carbon dioxide has no effect on these rates. As irradiance is increased further, carbon dioxide supply becomes limiting and the rate of photosynthesis approaches a plateau in which it is independent of the level of irradiance. The level of this plateau can be raised by increasing carbon dioxide or lowered by diminishing it. The rate of light-limited photosynthesis is normally independent of temperature, photochemical reactions being temperature insensitive, but when it is light-saturated it becomes temperature sensitive because it is now limited by ordinary chemical reactions. At high irradiances, approaching full sunlight, the rate of photosynthesis may fall.

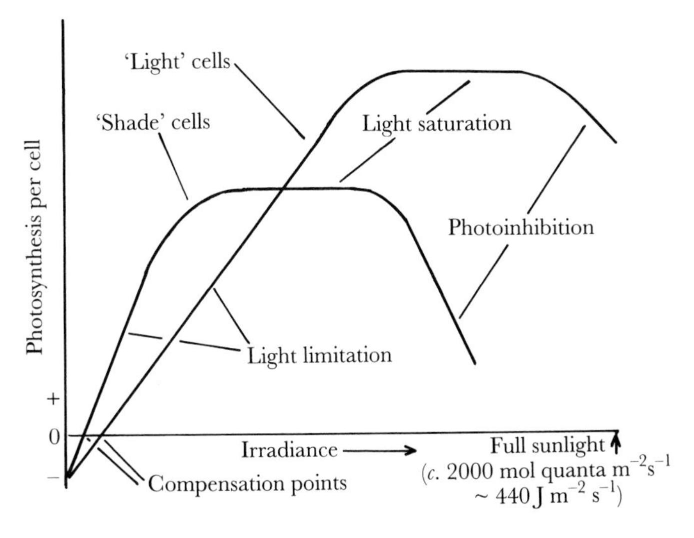

Fig. 2.6 Relation of rate of photosynthesis to irradiance for light- and dark-adapted cells. The relations are generally similar for both unicellular algae and multicellular plants.

This *photoinhibition* happens because the absorption of light energy exceeds the capacity of the chemical reactions of photosynthesis to utilize it and excess energy becomes diverted to destructive oxidative processes. The irradiances at which the compensation point, light saturation, and photoinhibition occur depend on the physiological state of the plant and particularly on the light levels to which it has previously been exposed. In *shade plants*, which usually contain more photosynthetic piments, these points occur at lower irradiances and the rate of photosynthesis per cell (but not per unit amount of chlorophyll) is higher at low irradiances than in *sun plants* (Fig. 2.6). Unicellular plants may adapt from one state to the other within the space of one cell division cycle, perhaps 24 hours, but more complex plants take longer. In polar plants the effects of low irradiance and low temperature are interrelated, both lowering the compensation point. The lower rate of respiration enables net photosynthesis and growth to take place at lower light levels than would otherwise be possible and by economizing on use of reserves enables the plant to survive for longer in darkness.

2.5.2 Ultraviolet radiation

The potential damage to biological systems from shorter wavelength UV (280–320 nm, referred to as UV-B) is illustrated by the estimate that radiation of wavelength 295 nm has a thousand times the potency in sunburning human skin as does the 320 nm wavelength. The impact on polar organisms depends on their level of organization and particular habitat and there are various general mechanisms for countering the challenge (Karentz 1991).

Damage to DNA may have mutagenic or lethal effects but repair is possible. The capacity to repair is genetically determined and microorganisms and human beings deficient in this respect are killed by low doses of UV that scarcely affect repair-efficient strains. A DNA repair-deficient strain of the bacterium *Escherichia coli* has been used as a biological dosimeter to measure, for example, the penetration of lethal radiation into the sea. Apart from genetic damage, absorption of UV can disrupt RNA, proteins (including enzymes and hormones), pigments, and other metabolically important molecules. Whether or not a cell is able to make good such damage and survive depends, among other things, on its physiological condition.

Protection is provided by substances that absorb UV without damage being caused. Such substances are the melanin produced by man, mycosporine amino acids contained in some Antarctic invertebrates and algae, and flavonoids in mosses and flowering plants. There is great variation, even between related species, in ability to produce these substances. Again, the ability is genetically determined. In animals, behavioural characteristics are important. Most animals avoid full sunlight as far as possible – only 'Mad dogs and Englishmen go out in the midday sun'. Marine birds and mammals that breed ashore may not be able to avoid it but are largely protected by feather or fur. Eyes are at risk but Antarctic penguins and skuas have higher thresholds for UV damage of the cornea than domesticated fowls. Tears of the southern sea elephant (*Mirounga leonina*) show strong absorption of wavelengths shorter than 300 nm. Many insect species have vision extending to 300 nm. Increase in radiation of less than 320 nm caused by reduction in the ozone shield would be perceived by them with possible effects on behaviour patterns.

It may be that increased exposure to UV-B will have little effect on the total biological productivity of polar habitats but because protection and repair are genetically determined there may well be selective effects on species in the course of time. Such changes should first be evident among microorganisms with short generation times.

2.6 Biological responses to long-term changes

At its maximum, during the Würm* glaciation some 20 000 BP, ice extended to a depth of up to 2000 m over the whole of Scandinavia, northern Russia and the northern European lowlands to meet an ice sheet covering the British Isles. North America was almost entirely glaciated as far south as Kentucky and its ice sheet probably joined with that of Greenland. The Antarctic ice sheet, on the other hand, being limited by attack from the surrounding ocean, was only slightly larger than it is at present. A simple view would be that marine plants and animals withdrew before the advancing ice and that, when it retreated, recolonization of habitats would not present any major problem to species which are mobile or have drifting propagules (but see pp. 49 and 50). On land, some plants and animals seem to have survived in favoured spots in both polar regions.

2.6.1 Survival in refugia

In the Arctic there is geological evidence, as well as that from the distribution of species, for numerous refugia remaining ice-free within glaciated areas at the time of the last glacial maximum. As the Pleistocene ice sheets spread south a zone of low precipitation developed on their northern edges with a consequent shrinking of the extent of the ice (p. 28). This, together with a lowering of sea level, created extensive potential refugia. The most important were in the interior of Alaska and Yukon, eastern Siberia, and a region known as Beringia, between the two which was exposed intermittently as sea level oscillated by 100 m or so. Beringia provided a corridor for migration of plants and animals from the unglaciated forest and mountain

*The Würm glaciation in the European Alps, the most recent, was contemporaneous with the Weichselian (Northern Europe), the Devensian (British Isles), the Wisconsin (North America), the Valdaian (Russia), and the most recent extension of Antarctic ice. The name 'Würm' is used here simply because it is the shortest.

areas of central Asia until it was finally inundated about 10 000 years ago. The vegetation which covered it consisted mainly of steppe-tundra, dominated by grasses and sagebrush (*Artemisia*), which supported large herds of grazing mammals such as bison, horse, and mammoth. Smaller and less hospitable refugia, provided by peaks protruding through the ice sheet – *nunataks* – enabled mosses, lichens, and a microflora of algae, yeasts, and bacteria to survive (Sage 1986). The Arctic poppy, *Papaver radicatum* (see Fig. 4.11), evidently survived the Würm glaciation in a refugium in northern Norway (Crawford 1995).

On continental Antarctica there would presumably be few refugia during the glacial maximum. The dry valleys remained unglaciated at this time but were more arid even than they are today and are unlikely to have supported a flora and fauna other than microorganisms. Nunataks remained above the ice and, since today they support mosses as far south as 84°S and lichens and cyanobacteria beyond 86°S (Kappen in Friedmann 1993), it seems reasonable to assume that they did so in the past. Among the less obvious organisms are bacteria and yeasts, recovered in a viable state from 87°21′S. The Polar Front must have been further north at the height of the glaciation and the peri-Antarctic islands seem to have been covered by ice caps extending out to sea. They may, of course, have had nunataks but Kerguelen was probably the only major refugium for preglacial fauna and flora.

Populations isolated in refugia evolve along their own lines. Both the Arctic and Antarctic have endemic species of algae, lichens, and mosses – forms most likely to survive on nunataks – but the number of these is few compared with those of cosmopolitan species. Endemic species, of course, can also arise by genetic change in relatively recent invaders. Bipolar species are also few. The two polar regions seem to have recruited most of their floras and faunas independently from their respective hemispheres, presumably mainly from alpine localities. Arctic species of flowering plants and animals frequently seem different from their counterparts in temperate climes but it is not always clear whether this is due to phenotypic plasticity or genetic change. The circumpolar Arctic-alpine bitter cress (*Cardamine bellidifolia*), transplanted from alpine habitats in Alaska, British Columbia or Yukon, into temperate conditions, grows into a tall plant. Seeds of the same species collected on Ellesmere Island (*c.* 80°7 80°W) give rise to dwarf plants under the same temperate conditions, showing that they come from an ecotype genetically different from the others. Greenland freshwater cladoceran crustacea appear to afford another instance of evolution in refugia. One of the most widely distributed of these today in the Arctic is *Chydorus arcticus*. Remains of a form similar to, but not identical with, this species and the closely related *C. sphaericus* have been found in 1.5 million-year-old Pliocene/Pleistocene deposits in northern Greenland. Thus, species having Arctic distributions at the present time seem to have evolved in ice-free refugia and become better adapted to polar conditions (Røen 1994).

2.6.2 Colonization

2.6.2.1 Invasion of the Arctic

Organisms may also invade both newly formed and established habitats from elsewhere. At first sight there seems little problem about how this may happen in the Arctic. Land masses projecting northwards provide avenues along which microorganisms, plants, and the smaller animals can be carried by wind, water, or larger animals migrating in search of fresh pastures. Beringia provided a highway for such movements, which culminated in the human invasion of Arctic America and Greenland from a heartland in northeast Siberia. Nevertheless, some biogeographical features of the Arctic are difficult to explain. The North Atlantic islands, including Iceland and Greenland, have floras and faunas which contain no true endemics but are dominated by Eurasian species. These include bulky flightless insects which one would suppose to be incapable of reaching the islands across wide stretches of sea with no favourable prevailing winds or currents to assist them. There are various theories as to how they got there from northwest Europe. One is that colonization took place in short intervals during the retreat of ice when fresh melt water provided an innocuous medium for transport and surface currents were temporarily deflected in the necessary directions (Coope 1986).

Studies using techniques of molecular biology indicate that the modern Arctic flora has components which survived in refugia as well as others that migrated from the south after the last glaciation (Crawford 1995).

2.6.2.2 Invasion of the Antarctic

There is evidence of retreat and thinning of the ice sheet around the margins of Antarctica in recent times. Much rock at present exposed, bears scars caused by glacial action. The ice cover on peri-Antarctic islands has likewise contracted. Invasion and colonization of the few habitats made available faced greater obstacles than in the Arctic, as is shown by the poorer species diversity. The circumpolar ocean is today about 900 km wide at its narrowest and the set of the prevailing winds and surface currents in the latitudes into which the southern extremities of America, Africa, and Australasia protrude, is easterly with little southerly component. Tree trunks originating from South America, carried by the Circumpolar Current, are occasionally washed up on the shores of sub-Antarctic islands and may serve as agents for biological immigration (Smith 1985). Seeds of *Colobanthus* (p. 83) survive immersion in seawater for three weeks but transoceanic dispersal of *Nothofagus*, the dominant tree genus of South America, seems unlikely (Burkle and Pokras 1991).

Dispersal of small propagules by wind is more probable. Following storms, samples of air over Signy Island show 13–24 times greater concentrations of biological particles than usual. Even these are much less, around $3.3\,m^{-3}$, than those over southern England, where 12 500 m^{-3} is typical in summer (Marshall 1996). Balloons in the upper atmosphere circle the continent in only eight days, with transit times between islands, such as South Georgia and Marion (46°52′S 37°51′E), of about two days. The necessary southerly component to carry particles to Antarctica is provided by cyclones. Virus particles, bacteria, microalgae, protozoa, fungi, bryophytes, ferns, and lichens all have dispersal units of such a size as might be transported by wind. To arrive in a viable state they must withstand desiccation, low temperatures, and exposure to UV radiation. Laboratory experiments show that some representatives of these groups do possess considerable resistance to these potentially lethal conditions. Ice fields can be important as collecting areas for propagules, brought by wind, which are released on melting to newly exposed substrates in the vicinity. Positively charged particles, such as algal cells, are collected by snow banks, which have negatively charged surfaces (W. S. Benninghoff and A. S. Benninghoff 1985). Direct demonstration of establishment of species carried by wind has not yet been achieved but there is convincing circumstantial evidence that it happens. Viable algae have been recovered from air over Antarctica (Vincent 1988) and the presence in moss cushions on the South Shetlands (*c.* 62°S 60°W) of pollen grains of South American plants suggests that viable spores of similar size may arrive too (Kappen and Straka 1988). Algae and bryophytes, not of local provenance, have appeared spontaneously around fumaroles both on Deception Island (*c.* 63°S 60°30′W) and on continental Antarctica (Broady *et al.* 1987). Evidence from snow and ice cores shows that this drizzle of airborne propagules has been continuous for many thousand years. There is, however, no direct evidence for transport by wind of insects or other metazoan animals to Antarctica.

Migrant birds can transport microorganisms. Arctic tern (*Sterna paradisaea*), cape pigeon (*Deption capensis*), southern giant petrel (*Macronectes giganteus*), and south polar skua (*Catharacta maccormicki*) sometimes carry viable algae or protozoa on feet, feathers, or bill, or internally in gullet or faeces (Schlichting *et al.* 1978). Viable cyanobacteria, the green alga *Chlorella*, and protozoa have been recovered from the tail feathers of the Arctic tern, which migrates between the two polar regions. Man, in comparatively recent times, has brought in, both intentionally and unintentionally, a variety of animals, plants, and microorganisms. The ecological consequences of this are considered later (p. 223).

2.6.2.3 Colonization and establishment

The first necessity for a study of colonization is sound taxonomy. The genetic relationship between an organism at a possible source and the one in the habitat it is supposed to have colonized must be established beyond question. The techniques of molecular genetics are obviously helpful here. Routes of immigration must also be examined. More information is needed than just the direction of the prevailing wind – trajectories of air masses and the probabilities of active convection within them need to be known.

Identification of propagules from birds must be supplemented by information on feeding grounds, migration routes, and behaviour patterns of the birds themselves.

When viable propagules have arrived in a potential habitat there are still the difficulties of establishment. Southwood's analysis of reproductive success in relation to habitat has been applied by Walton (in Kerry and Hempel 1990) to the examination of Antarctic problems. The various factors, which may vary both in time and space, can be grouped according to their interactions into three matrices:

(1) 'being there' – factors such as distance, weather patterns, and timing, frequency, and volume of propagule production;
(2) 'favourableness' – the value of an available microhabitat for the development of a propagule in terms of factors such as surface roughness, patterns of availability of water, and nutrients, light, and temperature; and
(3) 'uncertainty' – the probability of the right propagule landing in the right microhabitat and actually establishing itself.

These three matrices might be combined into a further matrix giving colonization success in probability terms. To apply this approach properly to polar habitats requires quantitative data which at present are lacking. At this stage one can only guess that the probability of a viable propagule establishing itself in a given place is low. That a substratum may have the potential to be colonized but that this is not realized unless conditions such as temperature and moisture become favourable, is well shown by the appearance of species, not previously evident, when a patch is covered by a protective cloche. Establishment is the most sensitive stage. Once it has been accomplished the organism is able to withstand conditions under which establishment would be impossible. A species with great ecological tolerance may flourish in an area in which, were it eradicated, the chances of re-establishment might be remote. However, some species well adapted to their habitat, for example, snow algae (p. 58) and emperor penguins (p. 167), do readily re-establish from neighbouring areas if they are eradicated from one locality. Examples of successful establishment of migrants are the chironomid midge and enchytraeid worm which were both accidently introduced on to Signy Island in plant material and established themselves, surviving for 17 years. Both were capable of supercooling and seem to have been pre-adapted, by extension of genetically determined physiological mechanisms, for survival under much harsher conditions than those in their original habitat (Block *et al.* 1984). Among Arctic flowering plants, ecotypic diversity, as seen in the purple saxifrage, *Saxifraga oppositifolia* (p. 88), is viewed as a pre-adaptation to Arctic conditions (Crawford 1995).

2.6.2.4 Colonization of recently deglaciated areas

Retreating glaciers, of which there is an abundance at the present time, leave behind them debris which is invaded by microorganisms, plants, and animals and eventually develops into a soil. As one moves away from the glacier snout one passes along a *chronosequence*, in which the stages of colonization are laid out, approximately at least, in order of age. Propagules are derived mainly from the immediate surroundings, being brought in by wind and run-off, and large numbers, rather than specialized adaptations, are most important in assuring rapid establishment. Bacteria, fungi, algae, and cyanobacteria establish themselves in and on the ground surface and stabilize it. Mosses and lichens follow. On Signy Island, vegetative fragments of mosses, such as leaves and stem apices, act as propagules, the ice cap evidently being an important reservoir supplying them. Far more show initial growth than eventually survive. In favourable sub-Arctic and alpine situations, shrubs and trees are among the earliest arrivals and relatively stable forest may be established in 100 years. In the Eurasian tundra zone, stable communities take 800–1000 years to develop. For Svalbard and harsher olimates the times required may be 1000–1500 or even 9000 years. Recently glaciated terrain provides a unique opportunity for the study of colonization and succession but it has to be remembered that it differs from the Late Glacial situation in that the areas are small and surrounded by established vegetation (Matthews 1992).

3 Glacial habitats

3.1 Introduction

The rare human explorers and birds venturing over polar ice caps encounter each other with surprise and interest. Ice is essentially inimical to life. Its presence implies low temperatures and in its pure form it is hard, virtually inpenetrable by organisms, and devoid of nutrient content. Most importantly, water, in the form of ice, is unavailable and the energy required to melt it and make it available is more than poikilothermic metabolic activity can supply. Pigmented organisms, such as algae, produce melting in their immediate vicinity by absorption of radiation but this ceases abruptly when the sun goes down. However, 99% of the continental land mass of Antarctica and 28% of the land surface of the Arctic are permanently ice-covered and we know that these areas receive a constant supply of viable, wind-carried, propagules of various kinds of organisms. One may wonder, therefore, whether life has got any sort of foothold in this unpromising habitat. It will be convenient to consider also those habitats which are entirely bounded by ice but in which some liquid water is available.

3.2 The physical characteristics of snow and ice

Freshly fallen snow is crystalline, loosely packed, light and fluffy or granular. Its conversion to ice involves change in grain structure with recrystallization and the gradual elimination of trapped air with concomitant increase in density. Sublimation and recondensation and, if temperatures are around freezing point, melting and refreezing, may play a part. Compaction and bonding between granules leads to formation of old compressed snow, known as *nevé* or *firn*. Maximum packing of this type gives a density of around 0.55 g cm^{-3}. The depth at which this is reached depends on temperature but may be around 10 m. Below this, further densification takes place by deformational recrystallization of the firn grains. Firn remains permeable to air but there is gradual diminution of the interconnecting channels until at a density of about 0.83 g cm^{-3} the air spaces seal off to form isolated tubular bubbles and the ice becomes impermeable. This conversion to *glacier ice* happens at around 70 m depth. Finally, the bubbles become compressed still further, the air in them reaching a pressure of 200–300 bar (ice from a deep core makes an entertaining addition to a gin and tonic) and the density approaches 0.92 g cm^{-3} (Seligman 1980).

Since snow is a poor conductor of heat, diurnal variations in temperature rarely penetrate more than a metre below the surface and seasonal variations diminish rapidly with depth. At 10 m temperature is almost constant the whole year round at about the mean annual surface value. At the South Pole this is around −50.5 °C. Ice behaves as a viscous plastic material and flows outwards from the centre of an ice cap under its own weight. The profile of an ice cap tends to be parabolic and near the centre, where the gradient of the surface slope is small, the flow is minimal. Towards the edge slopes are steeper and rates of flow higher. Where mountains channel the ice into glaciers, velocities are greater and strains produce crevasses and ice falls.

A core taken from an undisturbed area of ice cap contains material in chronological sequence. Dating can be done by counting annual layers, distinguished by differences in grain size, or content of the oxygen isotope ^{18}O, by locating bands of ash which can be related to volcanic eruptions, or by calculations based on models of ice flow. The air entrapped in bubbles can be analysed for its content of carbon dioxide and other gases and isotope ratios in the ice can be used to derive an estimate of the mean atmospheric temperature when the ice was laid down. From determinations such as these on cores obtained in Greenland and in Antarctica a remarkable picture of climatic changes over the past 160000 years has been obtained (Lorius 1991; Lorius *et al.* in Drewry *et al.* 1992). The ice from the polar ice caps is pure, averaging between 1 and 2 mg dissolved matter per litre with a small content of solid particles, between 0.3 and 5 μm in diameter, among which, as well as mineral matter, microorganisms or their remains have been demonstrated by electron microscopy.

3.3 Microorganisms in ice

The second Byrd expedition (1933–35) isolated a bacterium from snow samples taken in remote parts of the Antarctic continent, never before visited by man. This was probably the variable spore-forming heterotroph, *Bacillus subtilis*. It seems unlikely that it could sustain active growth in the cold, dry, situations where it was found and presumably it was present as spores. Later workers have found, in both Arctic and Antarctic ice, microorganisms characteristic of the atmospheric flora of more temperate latitudes, but in very small numbers. The mite *Nanorchestes antarcticus*, which has been seen moving at temperatures as low as −11 °C, sometimes occurs in large numbers in Antarctic ice. All its life stages have been found in ice cores from the McLeod Glacier, Signy Island, and it is presumed they were transported there by wind. It seems unlikely that they could maintain existence for long on an ice substratum.

The availability of dateable samples from cores gives an opportunity of finding out how long microorganisms can remain dormant but alive in ice, which has been taken by Russian workers in the Antarctic. Obviously, stringent precautions against contamination during sampling are required. Aseptic sampling was achieved by melting out the interior of a section of ice core. Potato broth with yeast extract, a medium judged to be the most generally suitable, was used to elicit growth. The chance of finding viable organisms diminished with depth; for example, if 20% of samples from the top 100 m contained viable organisms, only 3% of those from 1500 to 2405 m gave positive results. Viable fungi, yeasts, and spore-forming and non spore-forming bacteria were recovered from these depths, showing survival for several thousand years is possible for species belonging to all these groups (Fig. 3.1). The yeasts included both psychrotolerant and psychrophilic forms. Mycelial fungi were the most resistant, a species of *Penicillium* being recovered from 651 m in ice estimated to be 38 600 years old. These fungi are psychrophilic and have ultrastructural features which may be related to their resistance. The non spore-forming bacteria, mainly strains of *Pseudomonas*, were both mesophilic and psychrophilic. These two different physiological types were found in different horizons of the ice, suggesting that they came from different sources, the mesophiles from temperate latitudes and the others being indigenous to the Antarctic. Spore-forming bacteria, *Bacillus* spp., predominated in the deeper samples and can evidently survive for at least 12 500 years. Actinomycetes have also been found, both in snow and deep in the ice. One, from a depth of 85 m, about 2200 years old, is believed to be a new species, *Nocardiopsis antarcticus*, endemic to Antarctica. This organism is remarkable for synthesizing high

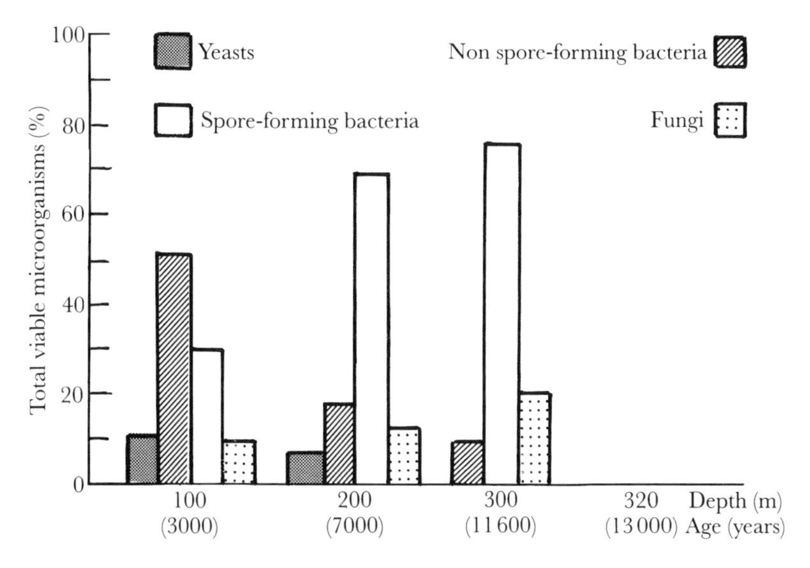

Fig. 3.1 Survival of microorganisms at different depths in the Antarctic ice sheet in terms of the numbers in different groups as percentages of the total of viable cells isolated. (Redrawn from Abyzov in Friedmann 1993, by courtesy of the author and John Wiley and Sons, Inc.)

intracellular concentrations of melanin, a UV-absorbing material (Abyzov in Friedmann 1993). No search for viable cyanobacteria or algae appears to have been made. There can be no doubt that all these organisms present in deep ice were in a dormant condition, because of absence of liquid water as much as low temperature.

3.4 Subglacial lakes

Liquid water is available in places beneath the Antarctic ice cap. In the region of the Pole of Inaccessibility (*c.* 83°S 63°E) radar soundings from the air showed changes in signal from one characteristic of an ice/rock interface to one indicative of an ice/water interface, suggesting lakes held between bedrock and ice of 3000–4000 m thickness. There may be as many as 70 such lakes. Geothermal heat warms the bottom of the ice and its thermal conductivity is low enough for this to bring the temperature to melting point, although at the surface it may be −60 °C. Given an energy source there seems no reason why life should not be possible in these lakes. Organic matter from the remains of organisms incorporated in the ice might support a low biomass at a low rate of turnover but getting a sample to see whether this is so without irretrievable contamination of a lake is an extremely tricky technological problem (Ellis-Evans and Wynn-Williams 1996).

3.5 Proglacial lakes

Lakes contained in depressions in an ice sheet exist in a few places. Lake Untersee, at 71°20′S 13°26′E in the Wohlthat Massif, is permanently contained under 3 m of ice and seems to have been ice-covered since it was formed. It is partially bounded by a glacier which feeds the lake by underwater melting. There is no outflow and water is lost only by ablation of the ice. The lake water is alkaline, pH 10.5, and its chloride concentration, *c.* 45 $g\,m^{-3}$, greatly exceeds that in precipitation, 0.05–1 $g\,m^{-3}$, suggesting that the present lake is a remnant of a much larger one that has been reduced in volume by ablation over a long period, perhaps going back to the Holocene climatic optimum 5000 BP. Temperature and salinity are uniform down to 79 m because of convectional mixing but a trough going down to 105 m contains stratified, anoxic, water. The conditions of light, temperature, and nutrient availability are such that plant growth is possible but primary productivity is low, 0.1–1.1 $\mu g\,C\,l^{-1}\,day^{-1}$. The bottom sediment of the lake, several millimetres in thickness, is of an organic nature and seems to contain methane (Priddle and Heywood 1980; Wand *et al.* 1997).

3.6 Life at the ice margin

Where the ice cap comes into proximity with low albedo rock, there is the possibility of local warming and melting of ice and snow providing environments where active life is possible. These take the form of melt holes or surface pools filled with melt water. Snow can provide an ephemeral habitat in which organisms surrounded by ice crystals may find sufficient moisture and other necessaries for growth. Where land ice extends over the sea there may also be situations in which freshwater life is able to establish itself.

3.6.1 Cryoconite holes

Cryoconite is the rock debris which is blown on to ice surfaces. It absorbs radiation and melts the underlying ice to produce a hole with near vertical parallel sides with the debris lying at the bottom. Once formed, the development of the hole may be speeded because it collects more debris and because the relatively warm water flowing over the ice surface drains into it. Water has a higher albedo and higher thermal capacity than ice and the organisms which may colonize the hole absorb solar radiation and so may promote its enlargement. Usually, cryoconite holes are a few tens of centimetres deep and wide but they may join up and form pools several metres across.

The water in the hole contains small amounts of mineral nutrients derived from the rock dust and may be inoculated with wind-carried propagules or pieces of algal mat from nearby lakes or streams (p. 125). Cryoconite holes on the Canada Glacier in Victoria Land (77°38′S 162°53′E) contain algal mats of which the main component is *Phormidium frigidum*, accompanied by other cyanobacteria, some of which, such as *Nostoc punctiforme* are nitrogen-fixing and therefore able to contribute to the productivity of the microhabitat. Embedded in the mat there may be cells of the green alga *Tetracystis*, often with an orange colour due to carotenoids. Occasionally, flagellates and diatoms are present (Vincent 1988). Diatoms have also been reported from cryoconite holes in the northern hemisphere. There seem to be no accounts of the fauna of cryoconite holes but one would expect to find in them protozoa, rotifers, tardigrades, and nematodes, such as are associated with algal growths in the same general areas.

3.6.2 Supraglacial pools

Cryoconite holes may enlarge or melt water may be impounded to form substantial pools on an ice surface. On the McMurdo Ice Shelf at 78°00′S 165°35′E, an area of gently undulating ice covered with a sediment of marine origin, has numerous pools and temporary streams (Vincent 1988; James *et al.* 1995). While these may persist for decades, the habitat is essentially ephemeral, changing as the ice shelf moves. Usually the pools contain thick mats of cyanobacteria, pink or grey in colour and generally similar to those in cryoconite holes (Fig. 3.2). Some have *Phormidium* spp. as the dominant forms, others *Nostoc punctiforme*, and diatoms are present. Analyses of these mats shows that much of the dry weight is contributed by mineral matter and that the content of nitrogen relative to both carbon and phosphorus is low, although the mats do have some nitrogen-fixing

Fig. 3.2 A small melt pool on the McMurdo Ice Shelf showing the benthic mat of cyanobacteria. (From Vincent 1988, by courtesy of the author and Cambridge University Press.)

capacity. The chlorophyll *a* content is low, with a ratio to carbon of less than 0.01 : 1, and of this much is photosynthetically inactive. This suggests that decomposition of the algal material is slow, which may be attributable to low temperature suppression of heterotrophic activity and detritivore grazing. Whereas small bactivorous ciliates are abundant, large herbivorous ones are lacking. Furthermore, of course, these waters are frozen for much of the time. The ecology and physiology of the mats seems similar to those in shallow waters on adjacent soil and rock (pp. 118 and 131). Planktonic algae form only a small proportion of the total biomass in these pools, a varied assemblage of species being found with heterotrophic cryptophytes tending to dominate.

The freezing temperatures, winds, and low relative humidities, to which these communities are exposed, cause prolonged periods of severe desiccation. Stream communities are particularly liable to rapidly alternating wetting and drying whereas deep ponds may retain some liquid water for much of the year. The rapid recovery of algal mats after desiccation has already been noted (p. 42). Viable cells may be still present in *Phormidium* mats after three years in the dry condition.

3.6.3 The snow alga community

Snow fields in all parts of the world, in alpine regions as well as in the maritime Arctic and Antarctic, often develop patches of colour where ablation is occurring. The patches, usually red, but sometimes green, yellow, or grey, are striking. A cartoon by George Cruikshank, depicting John Ross's return from his Northwest Passage expedition, entitled 'Landing the Treasures, or Results of the Polar Expedition!!! at Whitehall 17 Dec. 1818' shows a barrel labelled 'Red snow 4 BM [for British Museum]' as its centrepiece.

The red colour is due to the presence of spores of green algae, such as *Chlamydomonas nivalis* and *Chlorosphaera antarctica*, which are coloured intensely red by accumulation of carotenoids. Green snow, which occurs where snow becomes waterlogged by drainage from above and the availability of combined nitrogen is higher, contains other green algae, such as *Hormidium subtile* and *Raphidonema nivale*. Yellow snow may contain chlorotic green algae or chrysophytes. Often grey snow is that colour because of rock dust but, sometimes, it contains a desmid, *Mesotaenium berggrenii*, a green alga with a reddish-brown pigment in its cell sap. Bacteria and fungi occur in snow but have been little studied. 'Snow fleas', mainly small wingless insects of the genus *Podura*, appearing on the top of snow or ice, are well known in the Arctic. These, however, are soil-dwelling and work their way up to the snow surface when conditions get warmer in the spring. They are not in the same category as snow algae.

The primary requirement for growth of the algae is liquid water and they are usually found where the mean air temperature reaches 0 °C or more for a period of about a month. Nevertheless, they have been occasionally seen in southern Victoria Land (*c.* 77°S 160°E), where even in summer the mean air temperatures lie several degrees below zero. Being pigmented, the algae absorb radiant energy thus achieving a higher temperature than the surrounding snow and some liquid water for growth. Since 1% of the incident radiation may penetrate 1 m into snow, this may occur at depth. Dust and other impurities from the air provide the necessary nutrients. Snow on Signy Island, which has extensive areas of snow algae, contains between 0.38 and 1.09 $mg\,l^{-1}$ of calcium ion and 0.56–2.48 $mg\,l^{-1}$ of magnesium ion, derived from sea spray. The presence of penguin and seal colonies on the island ensures seasonal supply of ammonia via the air. On islands in the vicinity of Palmer Station (64°46′S 64°03′W) total combined nitrogen concentration in snow containing algae ranges between 1 and 13 $mg\,l^{-1}$. In full sun the intensity of photosynthetically active and UV radiation in the surface layer of snow may reach damaging levels. The carotenoids in the red forms and the pigment in *Mesotaenium* afford protection against this damage. The carotenoid astaxanthin, which is not concerned in photosynthesis, is accumulated in *Chlamydomonas nivalis* outside the chloroplasts in lipid globules in response to nitrogen deficiency and its presence reduces the amount of potentially damaging radiation reaching the chloroplasts. In experiments with snow algae contained in UV-transmitting quartz vessels, UV radiation was found to inhibit the photosynthesis of red snow by about 25%, whereas in green snow the inhibition was 85% (Thomas and Duval 1995).

Various processes, both active and passive, affect the distribution of algae through the depth of snow. Some species have motile stages and disperse by swimming in water films around snow crystals. On sunny days patches may not be so obvious as on dull days. A possible mechanism for this apparent avoidance of high irradiation is based on experiments with *Chlamydomonas nivalis* at Finse in Norway (60°36′N 7°30′E). Counts at different depths showed that whereas in less intense light cells were held in the water films surrounding ice crystals, in sunlight the association with the water films decreased and they were carried down into the snow by melt water (Grinde 1993). During periods of ablation patches of algae may appear within a few hours, giving an impression of rapid growth. This is illusory; rates of photosynthesis of snow algae measured *in situ* are only sufficient to provide for a doubling in cell number in around 23 days. Primary productivity is correspondingly low, about $10\,\mathrm{mg\,C\,m^{-2}\,day^{-1}}$, under favourable conditions. Examination of the vertical distribution of cells shows that after a snow fall

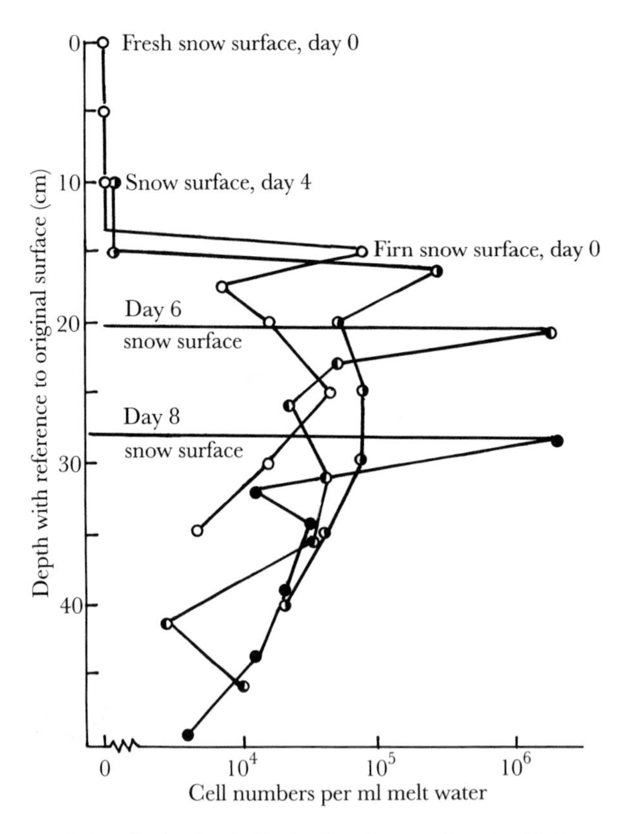

Fig. 3.3 Changes over 8 days in the depth distribution of snow algae in ablating snow, Signy Island. Cell numbers are plotted on a logarithmic scale. During the period of observation the peak concentrations of cells increased 20-fold although the total number of cells in the snow column rose by only 5-fold. (Redrawn from Fogg 1967, courtesy of the Royal Society of London.)

the peak concentration remains at the top of the firn, beneath the fresh snow, and below this numbers are small. As the snow ablates, the layer containing the peak concentration is exposed and with further ablation it remains at the surface, collecting up the cells in the snow below as it sinks. The development of a patch is thus more a matter of concentration of cells already present than multiplication to produce new ones (Fig. 3.3). Numbers of bacteria are greater, up to $6 \times 10^5 ml^{-1}$, and bacterial production is higher, in red snow than in white snow. Snow algae release about 10% of their photosynthetic production in extracellular form and this seems sufficient to provide for the bacterial growth observed (Thomas and Duval 1995). It seems, therefore, that we are dealing here with a community having both phototrophic and heterotrophic components.

Like the communities of cryoconite holes and supraglacial pools, snow algae patches are ephemeral, liable to be destroyed abruptly by ice movements or change in weather pattern. If the snow ablates completely, their cells, which tend to agglutinate, are left on the substratum to desiccate and be distributed by wind. Because of their wide distribution snow algae have potential importance on the global scale. The patches of low albedo, capable of growing and multiplying, which they produce, have appreciable local effect in accelerating disappearance of snow, although this does not often extend over a whole snow field (Thomas and Duval 1995).

4 Periglacial terrestrial habitats

4.1 Introduction

Leaving ice and snow we now turn to exposed rocks and soils as habitats, beginning with consideration of some of their physical and chemical characteristics.

4.2 Substrata

4.2.1 Exposed rock surfaces

A rock surface offers some hospitality to life, those organisms able to take advantage of it being termed *epilithic*. Much depends on aspect. On a clear day near the summer solstice the maximum radiation flux on a north-facing surface in the frigid Antarctic dry valleys is high, about $1050\,J\,m^{-2}\,s^{-1}$. If the surface is horizontal it is 650 and, if south-facing, $450\,J\,m^{-2}\,s^{-1}$. The temperature of the surface depends both on radiation and air temperature. During winter darkness, exposed rock surfaces are usually 1 or 2 °C below the ambient temperatures of −45 to −19 °C. In summer the temperature of a surface facing towards the equator may rise to as much as 42 °C above air temperature whereas one that is pole-facing may reach only 8 °C above. Although sloping surfaces reach higher peak temperatures, they are mostly at or below air temperature, having a diurnal range of up to 25 °C. Horizontal surfaces have a smaller variation of 10 to 15 °C (Nienow and Friedmann in Friedmann 1993). Snow and ice can ameliorate conditions. A thin ice sheet provides a warmer atmosphere below it, as in a greenhouse. Snow insulates against extreme temperatures.

The supply of water, arriving as precipitation, dew, frost, or run-off, is also affected by slope and aspect. Alternation between moist and dry conditions is least severe on horizontal surfaces. In coastal and maritime situations, melt water may trickle over rock surfaces for much of the summer. Crustose or foliose growth will retain water. Mineral nutrients may come in airborne dust, precipitation, run-off or splash, or, from the rock itself. Near bird colonies the nutrient supply may be so copious as to be toxic.

An overriding consideration is that organisms should be able to attach to the surface. If a rock weathers grain by grain or if it *exfoliates* (i.e. weathers by flaking of the surface), it may not provide a sufficiently permanent substrate, Different species may have preferences for different rocks. On Signy Island, quartz-mica-schist, an acidic rock, is colonized by the moss *Andreaea* and the lichens *Omphalodiscus* and *Usnea*, whereas adjacent marble, which is basic, supports the mosses *Tortula* spp. and the lichens *Caloplaca* and *Leptogium* (Fig. 4.1, Table 4.1). It is not clear whether texture or chemical composition is the determining factor here. On inland nunataks in continental Antarctica, lichens apparently only grow on rocks that have weathered for several thousand years (Kappen in Friedmann 1993).

Orientation relative to the prevailing wind is important, growth being favoured on the leeward side (Fig. 4.2). Abrasion by wind-blown rock debris or ice crystals may completely polish off any vegetation. Snow accumulating on the leeward side provides moisture whereas to windward evaporation is increased so that the time available for active metabolism is reduced. Wind speed decreases logarithmically as a surface is approached, the distance above the surface at which it becomes virtually zero being determined by an aerodynamic roughness factor, equivalent to about 0.001 cm over smooth ice, 0.1 cm for vegetation up to 1 cm high and 0.5 cm for smooth fresh snow.

Fig. 4.1 Junction between schist (on the left) and marble (on the right) Signy Island. The schist, which is acid, has *Usnea fasciata* and *Andreaea regularis* as dominant plants whereas the marble, basic, has *Leptogium puberulum*, *Tortula* spp., and *Grimmia antarctica*. (Photo G. E. Fogg.)

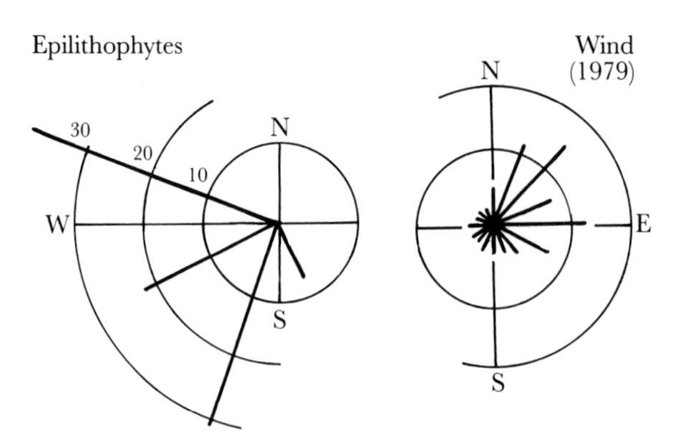

Fig. 4.2 Frequency of occurrence of epilithic cyanobacteria on rock surfaces with different aspects in relation to frequency of wind direction in the Vestfold Hills, Antarctica. (Redrawn from Broady in Friedmann 1993, by courtesy of the author and John Wiley and Sons, Inc.)

Thus, crustose or foliose growths reduce the effect of wind on convective heat transfer, evaporation, and direct mechanical damage.

Cracks in rocks may provide shelter for small invertebrates. In both the Arctic and Antarctic mites and collembola may often be abundant in such situations.

4.2.2 The endolithic habitat

Some lichens bore into rock but usually their bulk remains at the surface. In exfoliating rocks growth may take place beneath a plate and help its eventual detachment. Microorganisms may grow beneath the surface (i.e. are *endolithic*), in the cracks and fissures of weathered rocks (i.e. are *chasmolithic*), and others may colonize the interstices of porous rocks (i.e. are *cryptoendolithic*). In the dry valleys of Antarctica, quartz sandstone, with a grain size of 0.2–0.5 mm and a porosity of 8–15%, is the principal habitat for cryptoendolithic organisms. A silicified crust, about 0.1 mm thick, derived from wind-blown dust, forms on leeward faces of such rock. This case-hardening stabilizes the surface, reducing wind erosion and thereby giving time for colonization. Chasmoliths are found in the vertical fissures in quartz glacial debris in the Vestfold Hills (*c.* 68°35′S 78°10′E) and in cracks parallel to the surface in charnockite in the Mawson Rock area (67°36′S 62°57′E) (Nienow and Friedmann in Friedmann 1993).

The organic matter needed for endolithic growth is produced by photosynthesis and hence the rock must be to some extent translucent so that light reaches the microorganisms. Quartz is basically transparent but within sandstone, light becomes rapidly attenuated by reflection and absorption, particularly when, as is usual, iron oxyhydroxides stain the rock brown. The presence of material absorbing in the visible spectrum favours growth in so far as it lowers the albedo, thereby increasing the heat input. In each millimetre of dry rock, 70–95% of photosynthetically active radiation is absorbed. Infiltration of water into the rock increases its transparency, decreasing the extinction coefficient by about 25%, as a result of alteration of the refractive index between the quartz grains and the medium in the pores. Large transparent grains, acting as windows, admit light to deeper layers, whereas zones of pigmented organisms may distort the smooth curve of exponential decrease in light intensity that theory predicts.

In highly desiccated areas such as dry valleys the endolithic habit has the great advantage of conserving water. Once wetted, sandstone retains moisture for days or even weeks. Within the rock the freezing point of water becomes depressed to as low as −5 °C as a result of vapour pressure lowering at menisci in the capillaries of the micropores. It is significant that the endolithic communities of frigid and hot deserts are remarkably similar in composition and functions, desiccation overriding any effects of temperature.

4.2.3 Rock debris

Passing from warm to cold regions of the earth, chemical weathering of rocks becomes less important and physical processes bringing about decay predominate. These include glacial action, the effects of water in the forms of vapour, liquid, and ice, salt weathering, insolation, and wind action (Campbell and Claridge 1987). Glacial action is a major agent in the breaking and grinding of rock to rock flour and fragments, large and small. Cold glaciers, frozen to their beds, as found in Greenland and continental Antarctica, have little erosive power. Warm glaciers, at the pressure-melting temperature of ice, slide over their beds more easily and are more destructive. This type of glacier, which occurs in the less frigid parts of the Arctic and on Antarctic islands, produces more glacial *till* (unsorted debris) from which big moraines are built. With processes other than glacial action the form of product depends more on the kind of rock. Granite and sandstones crumble into crystals or grains, dolerite at first forms large prismatic blocks, foliate rocks such as shales, gneiss, and schist, split into plates (Fig. 4.3). In polar regions, the role of water in the disintegration of rocks is variable and mostly, as at high altitudes and in dry valleys, slight, Water, entering fine cracks, fracture planes, or pores, expands on freezing, breaking the rock. Crystallization of salts from solution held in cracks and pores has a similar effect. Solar radiation can be a powerful agent in disintegrating rock. Relatively high temperatures can be reached and cooling is rapid, resulting in stresses which can crack rocks. The effects are greater if water or salts are present. In some situations freeze–thaw cycles can exceed 100 per year. Wind carrying ice – which becomes as hard as window glass round about −50 °C – or sand, abrades rock and carries away material up to the size of small pebbles.

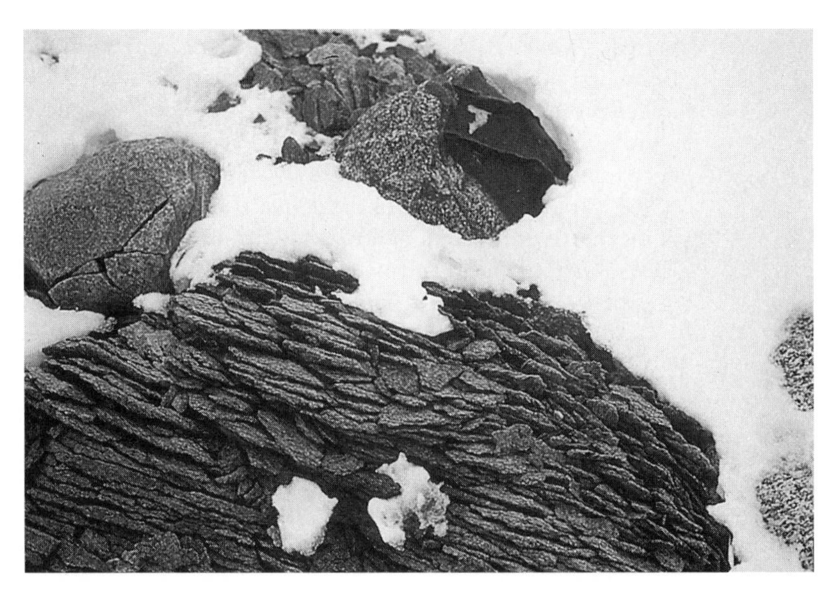

Fig. 4.3 Frost-shattered rocks, Stonington Island, 68°12′S 67°00′W. (Photo G. E. Fogg.)

There are extensive aeolian sands near the Colville River in Alaska (*c.* 69°30′N 152°W) and dunes of alluvial sand occur in the lower Victoria Valley in Antarctica (*c.* 77°20′S 162°E). Interbedded snow cements sand dunes against high winds. Rock detritus and soils can likewise be stabilized at depth by permafrost (p. 68). The surface of detritus may be too mobile for anything but temporary opportunistic colonization by plants. Screes move under gravity, flushes of melt water, or wind, may shift finer material, and movement caused by frost may be extensive. The effects of this last agency are commonly seen in high alpine as well as polar regions (Sage 1986; Campbell and Claridge 1987; Pielou 1994). Alternate freezing and thawing in the top few centimetres of a deposit is accompanied by expansion and contraction which result in 'frost boils' and stirring and sorting giving rise to patterned ground (Fig. 4.4). This probably starts with narrow vertical cracks forming a honeycomb pattern. If these remain open when thaw occurs, water drains into them and a succession of freeze–thaw cycles will build-up a wedge-shaped mass of ice. Ice wedges may be several metres across but, since the increment is only 0.5–1 mm a year, these must be thousands of years old. The polygons or circles thus demarcated range from less than a metre in the more frigid climates to 30 metres or more in sub-polar regions. The area enclosed is covered with fine material with the larger stones tending to accumulate around the edge. This sorting occurs because large stones are heaved up during freezing to be left in that position when the finer material is washed downwards by melt water. Because drainage is better around the perimeter the sludge in the centre of the polygon is

Fig. 4.4 Patterned ground, stone polygons (15 cm rule gives scale), Kilpisjärvi, 69°03′N 20°50′E. (Photo G. E. Fogg.)

higher, so that the surface is domed and the larger stones work their way down the slope to the edges. The continual movement is inimical to plant growth but stones give the perimeter sufficient stability for mosses and lichens to establish themselves. The insulation provided by this vegetation retards freezing around the perimeter and promotes the doming effect (Fig. 4.5). Polygonal patterned ground develops to a greater extent in the Arctic than in the Antarctic. If the site is sloping, polygons are replaced by parallel stone stripes running downhill. The flow is most rapid in the fine material in the centre and larger stones are pushed to the edge of the stripe. This is a special case of the general phenomenon of soil creep or *solifluction* – the slow but continual downhill movement of surface soil or rock fragments saturated with water.

Screes may remain stable long enough for vegetation to establish itself. The tops of stones may be occupied by epilithic lichens whereas deeper in the spaces between them there may be green algae adapted to lower light intensities. The undersides of translucent quartz pebbles lying on silt or soil may be colonized by *sublithic* (or *hypolithic*) green algae or cyanobacteria. Pebbles up to 80 mm thick may transmit 2.7–0.6% of vertically incident sunlight, the ellipsoidal shape concentrating the rays to some extent. The situation is one that conserves moisture and reduces temperature fluctuations and in dry habitats it may support the major part of the biomass.

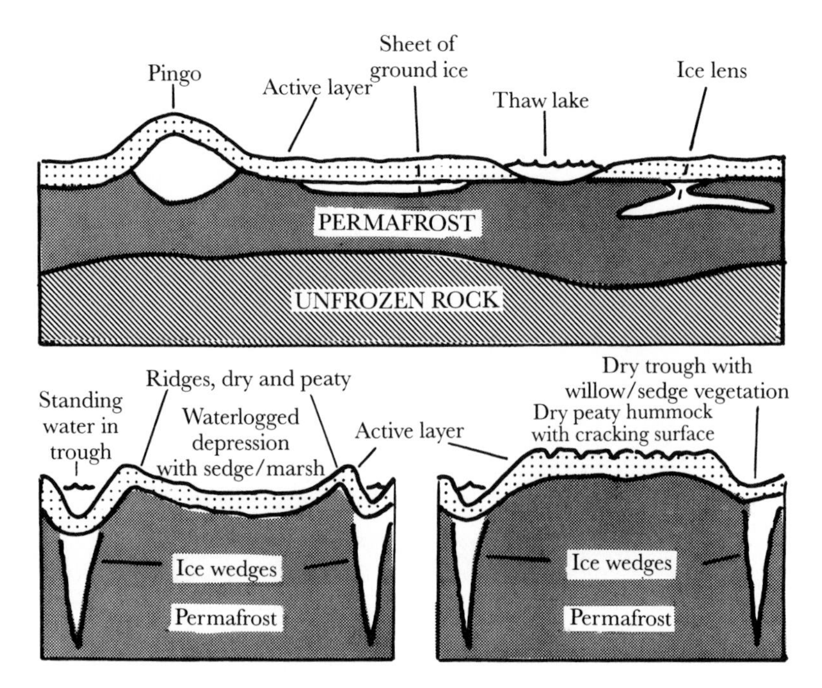

Fig. 4.5 Simplified sections, not drawn to scale, of surface features in Arctic tundra. (Modified after Sage 1986.)

4.2.4 Permafrost

Permafrost is earth material the temperature of which has been below 0 °C for several years so that the bedrock, clay, gravel, sand, silt, organic soil, whatever it may be, has become ice-cemented (Sage 1986, Fitzpatrick in Woodin and Marquiss 1997). Continuous permafrost covers some 7.6×10^6 km^2 of the northern hemisphere, with its southern limit, corresponding approximately with a mean annual air temperature of −8 °C, reaching 30 °N in parts of Eurasia. In places around the Arctic Ocean it extends under the sea. Its depth varies, being mostly between 200 and 700 m. Less is known of its distribution in Antarctica. Ground is frozen to a depth of 1000 m in places in the McMurdo Dry Valleys (*c.* 78°S 162°E) and probably to similar depths in other parts of the continent. It occurs on the South Shetland and South Orkney archipelagos but not on the sub-Antarctic islands.

Permafrost is of extraordinary toughness, hardly suitable for any form of life. However, indirectly, it has important effects in modifying overlying habitats. It forms an impermeable layer which plant roots cannot penetrate and which prevents drainage. In the Arctic it is nearly always associated with wet soils, having above it an 'active layer', perhaps 1 m deep but usually less, which freezes in winter but thaws in summer. Active layers are not present in the soils of the Transantarctic Mountains but do occur in coastal regions of the continent and on the peri-Antarctic islands.

Large masses of ice in permafrost affect surface topography and habitats (Fig. 4.5). The low vapour pressure of unfrozen water in the pore system of the active layer causes diffusion of water vapour from elsewhere and this freezes to form ice lenses. Massive ground ice gives rise to a dramatic permafrost feature, the *pingo*, a name derived from the Inuit word for conical hill. This is circular or oval in plan and 10–40 m or, rarely, 100 m in height. The core is a massive lump of ice originating from a silted up and shallowed lake. This is forced upwards by its own expansion during freezing and by pressure from the advancing permafrost around it. It can grow by as much as 25 cm per year by capture of moisture from its surroundings. Pingos are covered with soil and vegetation similar to those around them but the dome may be ruptured to form a crater with a pool in it. Pingos are characteristic of the Arctic, especially in the Mackenzie Delta area in Canada (*c.* 68°N 135°W). They are less common and well developed in the high Arctic and in the Antarctic, where ground water is generally less available.

Interactions between vegetation and permafrost are various and of importance both for the ecologist and civil engineer (Williams in Wadhams *et al.* 1995). On the whole, by maintaining low temperatures in summer, by promoting waterlogging, by restricting root growth, and by causing mechanical damage, permafrost inhibits plant growth. Vegetation, in turn, affects the depth at which permafrost occurs. The thicker the cover the greater is its insulating effect, the lower is the summer soil temperature, the nearer the surface is the permafrost, and the worse it is for deep-rooting plants. Bare soil surfaces, on the other hand, heat up more readily but are quickly colonized so that the ground again becomes insulated and colder.

4.2.5 Polar soils

With continued weathering and incorporation of organic matter, rock debris becomes soil but there is understandable doubt as to whether some of the surficial deposits of polar regions, with minimal biological components, deserve to be dignified by the name. However, even in the virtual absence of biological activity, processes of leaching, mineral decay, and segregation go on. The relations of their structure to time, topography, parent material, and climate give these deposits genetical connection with indisputable soils (Tedrow 1977; Campbell and Claridge 1987). Where glacial retreat is occurring the transformation of rock debris into soil is laid out in a chronosequence as one moves away from the glacier snout. In sub-Arctic regions the conversion of till into textured brown soil with developed horizons can be seen to take about 150 years (Matthews 1992).

Cold slows down chemical weathering and the rate of addition of organic matter. Aridity, too, slows down these two processes and retards segregation of materials by water movements. Modest availability of water permits weathering, some plant growth, and transport of soluble salts with separation into horizons. More water results in leaching of salts and, finally, waterlogging, which, by excluding oxygen, brings about accumulation of organic matter as peat. These various conditions and processes interact to produce from diverse starting materials a variety of different soils. Pedological classification is complex and the nomenclature horrific (Campbell and Claridge 1987, Fitzpatrick in Woodin and Marquiss 1997), but here a simple treatment will suffice. Tedrow (1977) distinguished four soil zones in polar regions – tundra, sub-polar desert, polar desert, and cold desert. The representation of these zones is vastly different in Arctic and Antarctic. In the Arctic, tundra and sub-polar desert extend over large areas, the polar desert zone includes all of the regions north of about 80°N, and cold desert is not represented. In the Antarctic, tundra is confined to the northern peri-Antarctic islands, sub-polar desert is absent, polar desert is restricted to limited maritime regions, and the rest of the continent is cold desert.

Cold, or *frigic*, desert soil, distinguished from other polar soils by the absence of organic horizons (but not necessarily by total absence of organic matter) is unique to Antarctica. The dry valley soils of Victoria Land are estimated to be about 4 million years old and for all that time have been subjected to continuous cold. In the Arctic, polar desert soils are only about 18 000 years old. Antarctic cold desert soils are strongly weathered and, being ahumic, water drains through them rapidly so that dryness is accentuated. In aridity, they are on a par with the soils of hot deserts, with which, indeed, they have features in common. These include granular disintegration and cavernous weathering of surface rocks, oxidation and staining of surface rock, endolithic algae, minimal weathering of buried rock, a crust at the surface and salt horizons within the profile, little organic matter, and being red or dark brown colour in the most weathered soils.

Polar desert soil, under less dry conditions, has ‘desert pavement’ – stones and pebbles – at its surface. The organic matter remains low but horizons are more prominent although soluble minerals are unleached. Under more moist but still cold conditions, as in the maritime Antarctic and northern Arctic coasts, chemical weathering and frost action are more evident, sorted polygons are formed, horizons are poorly developed and soluble minerals are leached. The organic content remains low. The mineral soil, having characteristic light grey colour and reducing properties, is of the type known as *gley*. Alongside these ahumic soils there may develop in more favoured spots, soil with an appreciable organic content, known as a *protoranker* or dry tundra soil (Table 4.1). In this, pads of humus develop beneath the isolated plant cushions. With more extensive moss or lichen cover the organic content may rise to 10%. The insulation provided by the cover allows surface temperatures to rise, promoting chemical weathering. On poorly drained sites there may be accumulation of peat-like material under moss turf (Table 4.1). This layer is often cemented by ice crystals and there is only slight penetration of humic material into the mineral soil below.

In the sub-polar zone, where higher plants with root systems penetrating many centimetres deep can grow, organic matter is distributed more

Table 4.1 Some chemical characteristics of polar soils

Type of soil and location	pH	Total organic C (%)	Total N (%)	C/N	Available nutrients (ppm)				
					P	Na	K	Ca	Mg
Dry valley soils, Victoria Land, Antarctica (1)	7.4 –8.9	0.02–0.09	0.004–0.085	6–13	0.03–0.27	8–1150	1–245	8–190	1–100
Cold-polar desert, Inglefield Land, Greenland (1)	8.8	0.2	–	–	–	2.3	11.7	15	3.7
Schist protoranker, Signy Is, Antarctica (2)	5.4	2.3	0.27	8.5	4	13	7	95	28
Marble protoranker, Signy Is, Antarctica (2)	7.9	2	0.26	7.7	8	24	5	596	17
Peat under moss turf, Signy Is, Antarctica (2)	4.7	43.6	1.36	32	3	102	17	213	196
Peat, sedge meadow, Devon Is, Arctic (2)	6.2	42.2	2.68	16	0	25	47	–	–
Brown earth under grass, Signy Is, Antarctica (2)	5.4	13.8	1.11	12.4	8	47	16	96	58
Brown earth, cushion plants, lichens, Devon Is, Arctic (2)	7.8	3.7	0.12	31	1	1	12	–	–
Ornithogenic, penguins, Signy Is, Antarctica (2)	6.1	10	1.8	5.5	460	69	73	106	68
Elephant seal wallow, Signy Is Antarctica (2)	6.7	30	3.58	8.4	66	199	100	220	134

Data from (1) Cameron 1969; (2) Rosswall and Heal (1975).

deeply and a loamy texture may result. This leads to true organic soils, brown soils, with developed horizons for which, in addition to rooting plants, active soil processes are necessary. Such soils are not found in the Antarctic. Soil mixing organisms, such as earthworms, are absent from the brown soils found on the southerly peri-Antarctic islands and in the more northerly parts of the Arctic but do occur on sub-Antarctic islands.

On the fringe of the Arctic in the tundra zone more mature brown soils, although frozen for part of the year, support vigorous root growth with earthworm activity. Brown soils grade into waterlogged peat soils in which addition of plant biomass exceeds its decomposition. Waterlogging, resulting in anoxic conditions, low temperatures, and acidity, favours peat accumulation. Under cold-polar conditions peats may be formed from the substantial algal growths in pools and streams (p. 120). In cool-polar zones, peats are produced, usually from mosses, some species of which are particularly resistant to decay, in soils overlying permafrost. Peat lands are extensive in the sub-Arctic but in the south largely restricted to the sub-Antarctic islands (see also p. 99).

4.3 Communities

Polar vegetation zones have been distinguished in various ways, none of which are entirely satisfactory. Here, the scheme of Longton (1988) will be used, as having the merits of simplicity and correlation of Arctic and Antarctic types, to provide a framework for accounts of both plant and animal communities (Figs 4.6, 4.7).

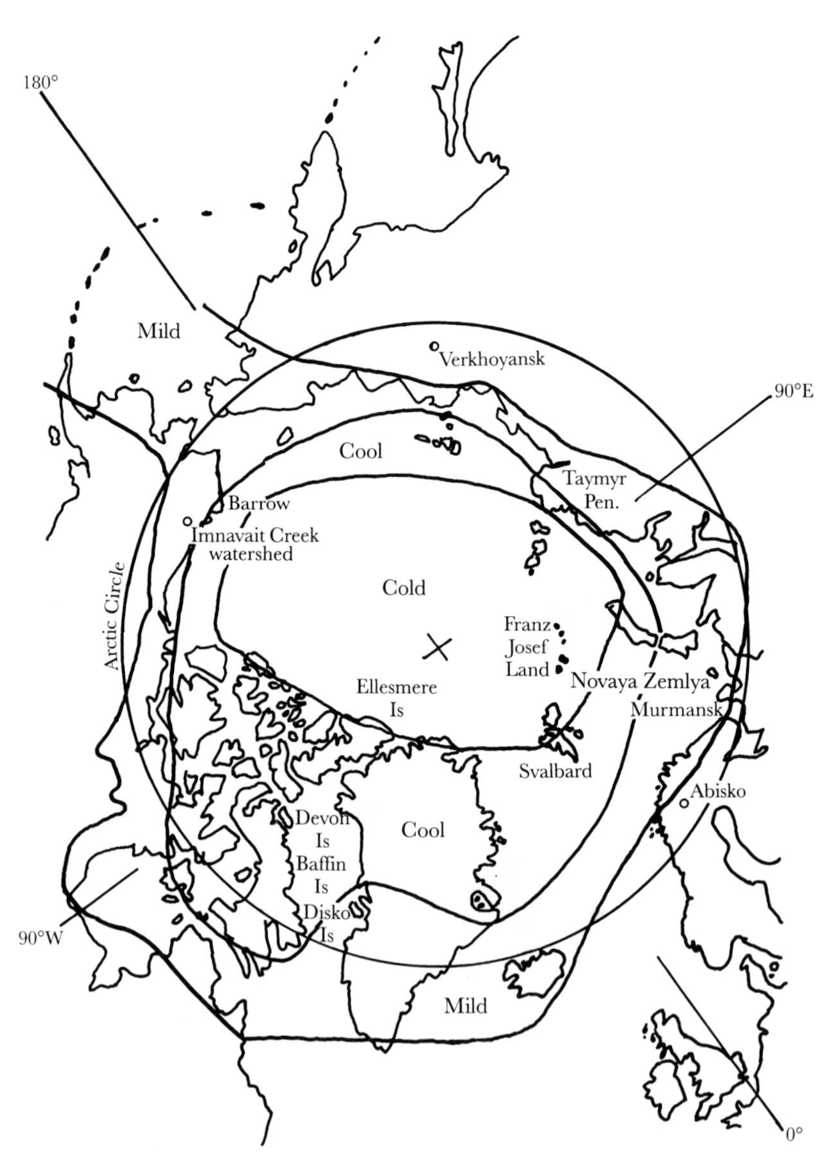

Fig. 4.6 Vegetation zones in the Arctic with some of the principal localities which have been studied. (Redrawn from Longton 1988, by courtesy of the author and Cambridge University Press.)

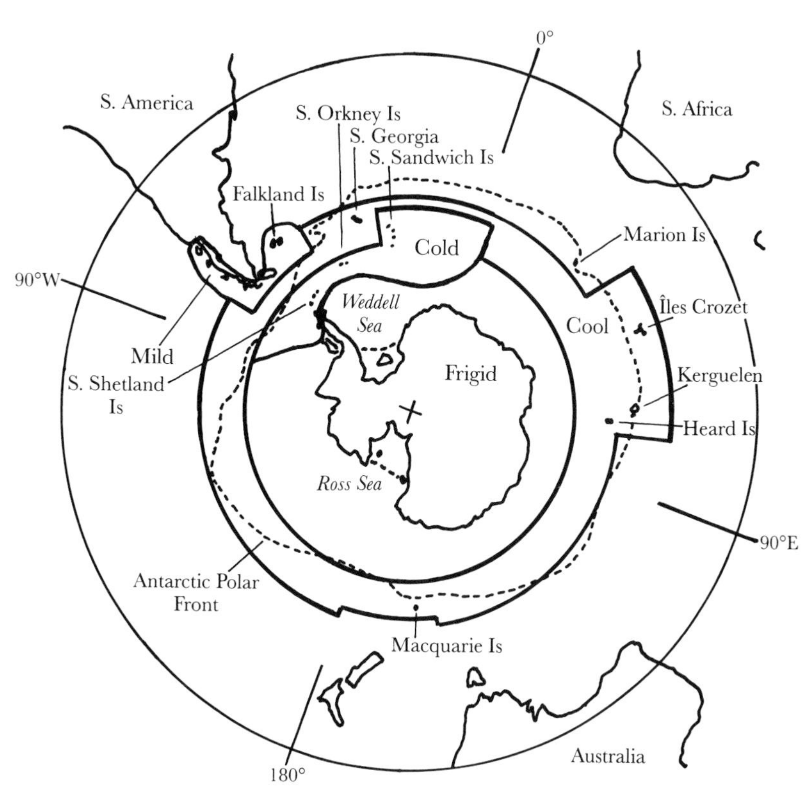

Fig. 4.7 Vegetation zones in the Antarctic with some of the principal peri-Antarctic islands which have been studied. (Redrawn from Longton 1988, by courtesy of the author and Cambridge University Press.)

Passing through these zones equatorwards from the poles one finds a general increase in productivity and species diversity but transitions are not smooth nor definite, because local features of climate, geology, and topography superimpose their effects. Communities are found out of place. Polar desert, for example, may occur on high ground in the cool-polar zone as well as in the cold-polar zone. Broadly speaking, communities of plants and animals are correlated with the soil types just described but seabirds and mammals select their breeding sites on land for accessibility and terrain, rather than soil properties. In the confusing mosaic of vegetation which is actually encountered plant sociologists have endeavoured to bring order by distinguishing formations, associations, sociations, and assemblages. For present purposes such refinements are not needed, the term 'community' having a convenient elasticity.

The successions in time of communities and soils are interwoven. Where conditions are inimical to plant growth, the substratum changes extremely slowly and remains much like its parent material. Vegetation quickens soil formation and modifies microclimate and hydrology. Under favourable conditions biotic control may be established and succession then tends towards the same type of community, the 'climax' (e.g. forest in the sub-Arctic), whatever the starting point may have been. In between these extremes succession is slower and divergent, giving rise to several different types of habitat, with communities, such as heath and lush meadow in the tundra, close together with steep environmental gradients between them.

4.3.1 The communities of the frigid-polar zone

The desert of the frigid zone of Antarctica occupies some 4000 km^2 of the McMurdo Dry Valleys region and lesser areas in the Bunger Hills, Vestfold Hills, and Peninsula. It presents a range of habitats from those in which extreme aridity makes life scarcely sustainable to others in which restricted communities are able to exist. Sheltered faces of bare rock may have lichens, such as *Buellia frigida*, even at relatively high altitudes (about 600 m) where cloud cover and summer snow flurries supply occasional moisture. A general survey of soil biota was made in southern Victoria Land using microbiological techniques based on those used in hot deserts (Cameron *et al.* 1970). Numbers of microorganisms per gram of soil vary between virtually zero and 10^7, as many being found at depths down to the permafrost as at the surface. The numbers of bacteria and protozoa vary with moisture content. Nematodes are distributed patchily, presumably in relation to these food organisms. Experiments with radiocarbon-labelled carbon dioxide show slight but significant photosynthetic activity associated with algae. Species diversity increases as numbers per gram of soil increase. Abundance and diversity are greatest in valleys with a north–south orientation. Slope, drainage, and exposure giving maximum duration, frequency, and quantity of insolation, available moisture, and protection from wind, are the favouring factors. There are high and low pH values, and high salt contents in the soil, which are inimical to microbial activity. Progressing from harsh to more favourable environments there is a sequence of different kinds of organisms. This is generally: (1) aerobic, heterotrophic, non-pigmented bacteria; (2) microaerophilic (i.e. preferring low oxygen concentrations), heterotrophic, pigmented bacteria; (3) actinomycetes; (4) coccoid green algae and both coccoid and filamentous cyanobacteria; (5) moulds, yeasts, and protozoa; (6) lichens; (7) mosses, with filamentous green algae, diatoms and nitrogen-fixing cyanobacteria. These various organisms are dormant for most of the time. Heterotrophic bacteria can find little in the way of substrates for active metabolism in the absence of phototrophic organisms. Activity by photosynthetic algae, lichens, and mosses is confined to brief periods with moisture and favourable temperature. The assemblage of species in any one locality seems a matter of chance colonization by airborne propagules. It is uncertain whether any of the bacteria are endemic and the microfungi almost certainly are not. On the other hand, *Cryptococcus vishniacii*, a psychrophilic yeast, seems to be endemic, having been sought, but not found, elsewhere in desert soils.

Hypolithic communities occur under pebbles of translucent quartz, the dominant forms being cyanobacteria.

In contrast to the random opportunistic nature of the unconfined flora of frigid zone desert soils, the endolithic communities with which they coexist have evolved to fit a particular niche. A broad programme of research on these communities in the McMurdo area has been pursued by Florida State University (Nienow and Friedmann in Friedmann 1993). Among cryptoendolithic communities of various types, all occurring in sandstone, the commonest is lichen-dominated and in fracture normal to the surface shows as distinct, parallel, coloured bands reaching about 10 mm in depth (Fig. 4.8). Within this depth the brown iron oxyhydroxide stain is absent, seemingly being dissolved by oxalic acid produced by the lichen. Typically, the layer just under the crust is black and next is a white zone. These contain filamentous fungi, the hyphae of which are black near the surface and white further down. Since they do not often produce reproductive structures these *mycobionts* are difficult to identify, especially since, in conforming with the pore system, they lose their characteristic morphology. Where the microhabitat is favourable the lichen may appear on the rock surface in a taxonomically recognizable form with reproductive structures. The green algae, the *phycobionts*, accompanying them include *Trebouxia* spp. Lower down is a green band, usually containing an endemic green alga, *Hemichloris*

Fig. 4.8 A section at right angles to a sandstone surface from a dry valley in Victoria Land, Antarctica, showing layered endolithic growth. (Courtesy of Dr D. Wynn-Williams and the British Antarctic Survey.)

antarctica, and sometimes below that is a blue-green zone of coccoid cyanobacteria. Other cryptoendolithic communities have *H. antarctica* or cyanobacteria, sometimes dark purple in colour, as the dominant organisms. Heterotrophic bacteria and yeasts are present together with all these phototrophs but their biomass amounts to less than 1% of the total. Chasmoendolithic microorganisms are found in low porosity sandstone, weathered dolerite, and granite, and, in contrast to cryptoendolithic communities, which are rare outside the dry valleys, are widespread in coastal regions of the continent. Epilithic lichens invade vertical cracks and in weathered granite extend horizontally beneath the surface. Coccoid cyanobacteria are widespread chasmoliths.

The physical conditions within these habitats have already been described (p. 64). The zonation is evidently the result of the radiation gradient within the rock although antibiotics, inhibiting algae and heterotrophic bacteria, may also be involved. Lichens, having higher light compensation points (p. 45) than free-living algae on account of the excess of heterotrophic biomass which has to be supported, need to be near the surface. The presence of *Hemichloris* in the lowest zone, where the photon flux is as low as 0.05–1 μmol quanta $m^{-2}s^{-1}$ when dry, and 2–10 μmol quanta $m^{-2}s^{-1}$ when wet, may be explained by this alga's sensitivity to high irradiance and its ability to achieve net photosynthesis at extremely low levels. Estimates of the total irradiation reaching the *Hemichloris* layer during a year suggest that the photosynthesis which can occur is sufficient only to support one or two divisions per cell.

Silica, the main component of sandstone, is transparent to UV radiation but iron-stained rock is opaque. Lichens produce UV-absorbing substances and the black layer in the cryptoendolithic community probably provides protection. Photo-oxidation caused by excess photosynthetically active radiation is seriously damaging to dry cells but repair mechanisms are effective when moisture permits active growth. Combined nitrogen appears to be in sufficient supply to maintain endolithic communities. Inorganic nitrogen as nitrate and ammonium salts is always present in low concentration. Biological nitrogen fixation is slight and the supply evidently comes in precipitation carrying nitrogen fixed in electrical discharges, manifest in the aurora, in the upper atmosphere. Other mineral nutrients are presumably mobilized from the rock itself by acids excreted by the microorganisms.

Estimates of the activity of endolithic communities have been made by measuring rates of carbon dioxide exchange (e.g. by using radiocarbon as a tracer or by infrared gas analysis). Uptake increases linearly up to about 200 μmol quanta $m^{-2}s^{-1}$, which is above the radiant flux likely to be encountered within the rock itself. Such measurements, made using crushed rock samples, in which exchange of gases is obviously much enhanced, are considerably higher than those made on intact communities. Carbon dioxide exchange at the rock surface gives a more realistic indication of what is happening. Net photosynthesis begins at a surface irradiation of about 100 μmol quanta $m^{-2}s^{-1}$ and increases up to 750 μmol quanta $m^{-2}s^{-1}$ without saturation. The lower temperature limit for measurable net carbon dioxide uptake is between −6 and −8 °C and the optimum between −3 and +6 °C with the upper compensation point between 8 and 15 °C. These responses are similar to those shown by lichens from sub-polar and alpine situations and do not indicate any special adaptation to frigid desert conditions. Under normal conditions optimum temperatures are rarely attained and then only on north-facing slopes. The maximum rates of net carbon dioxide uptake lie between 0.22 and 0.78 mg $CO_2\,m^{-2}h^{-1}$ as compared with an average value of 500 for temperate grassland under moist conditions.

Such information on the activities of endoliths in defined environments can be combined with long-term nanometeorological data to give estimates of primary production. From mathematical models of temperature and light regimes in the endolithic habitat the mean annual total time when metabolic activity is possible (i.e. temperatures above −10 °C and moist rock) has been estimated as 771 hours for horizontal rock and 421.5 for sloped rock (Fig. 4.9). This would allow gross community primary production amounting to 1215 mg C $m^{-2}yr^{-1}$ and net production of 606 mg C $m^{-2}yr^{-1}$. Dark respiration of organic carbon exceeds photosynthetic input under most conditions. These results do not match with the value of about 3 mg C $m^{-2}yr^{-1}$ for net ecosystem productivity derived from estimates of the average biomass of the community (about 30 mg m^{-2}) and its turn-over time (10 000 years). Admittedly these estimates are subject to uncertainties – values for biomass vary greatly according to the particular measure chosen

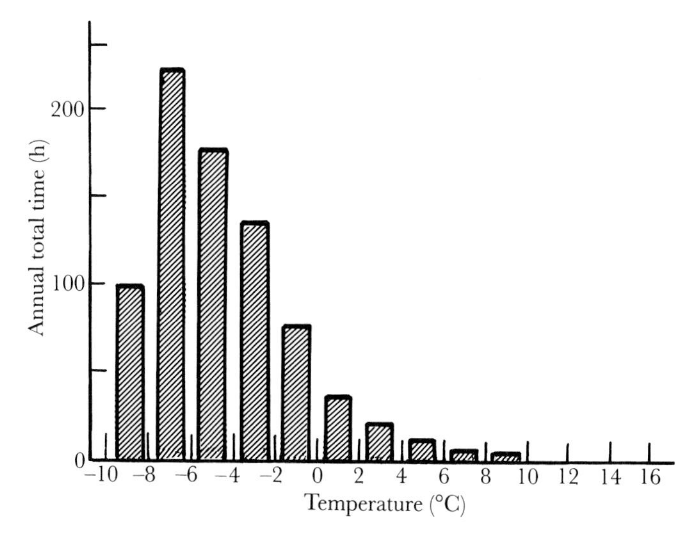

Fig. 4.9 Mean total times for each 2 °C interval during which metabolic activity of endolithic microorganisms is possible in horizontal rocks. (Redrawn from Friedmann 1993, by courtesy of the author and John Wiley and Sons, Inc.)

and estimation of turnover time in such heterogenous material is something of an act of faith – but it seems that less than 1% of the gross primary production may be incorporated into the standing biomass, the greater part of the primary production being lost as extracellular products. Frequent freeze–thaw cycles would cause release from cells (p. 39) and, in fact, various organic solutes have been detected in rocks harbouring endoliths. By acting as osmoregulators such substances might play a part in the survival of the communities.

Its low rate of gas exchange with the environment and the presence of both phototrophs and heterotrophs suggest that the cryptoendolithic community is more or less sealed in by the surface crust and self-contained. This conserves water and the age of the carbon in the communities indicates that they can survive successfully in this way for a thousand years or more. The primary colonization of the rock presumably takes place when cracks in the crust allow access of water and propagules. Development of the community and accompanying biogenous weathering then takes place parallel to the surface. Ultimately growth results in exfoliation and exposure with dispersion or destruction of the organisms. Biological and geological time scales overlap (Nienow and Friedmann in Friedmann 1993). On dispersion, the biomass contributes its mite to the soils of the high desert or valley floor. Throughout the life of the community there may be some leaching of organic matter and on exfoliation there will be more. This dissolved organic matter may percolate down through the soil to streams and lakes, being carried in the saline groundwater which flows along the top of the permafrost from higher elevations. This could provide a source of organic carbon for heterotrophic microbial life.

No habitat comparable in severity with the Antarctic frigid polar desert is to be found in the Arctic.

4.3.2 The communities of the cold-polar zones

Mean monthly air temperatures rising to 2 °C allow a greater abundance and variety of life in this zone. Dominance of mosses, together with *foliose* (flattened but not firmly attached to the substratum) and *fruticose* (strap-shaped or shrubby) lichens, and, in some regions, cushions of flowering plants, differentiates the cold-polar desert from that of the frigid zone. The vegetation remains scattered, with bare areas in between. It merges on the one hand with frigid-polar desert and on the other with tundra, existing where conditions are not so extremely cold and dry as in the former, but not sufficiently benign to support much growth of flowering plants. *Tundra* is a rather elastic term, often taken to include all polar vegetation beyond the tree line. Polar desert may thus be included as cold dry tundra.

The term *desert* is difficult to apply with exactitude in this zone. In temperate regions a rough approximation is that a desert has an annual precipitation of less than 250 mm, but much depends on how this is distributed over the year and on its balance with loss of water by evaporation. Measurement of precipitation in polar regions is complicated by snow drift, so that reported values are sometimes uncertain, but around the Arctic basin they are between 70 and 200 mm per year, with an average of about 135 mm. In Antarctica, for example, in the Bunger Hills (66°15′S 101°E) and Schirmacher Hills (70°45′S 11°40′E), which have some snow cover in the winter, values are higher. Around the tip of the Peninsula they may reach 400 mm. Reduction in temperature decreases the water-holding capacity of air and thus, if the actual water content remains the same, increases relative humidity. This will minimize evapo-transpiration so that what may be thought to be desert may be adequately moist.

Aleksandrova (1980) emphasized the importance of the combination of snow cover in winter with exposure in summer. Winter snow prevents erosion and damage by wind as well as providing insulation. Its insulating capacity depends mainly on the amount of air it contains. Fresh snow which has fallen as big flakes during calm and relatively warm weather conducts heat to a minimal extent. Compacted snow and firn are less efficient insulators. The difference in density between fresh and compacted snow is 10-fold and in heat conduction it is about a 100-fold. Melting snow is two or three times more conducting than dry snow of the same density. Snow cover allows soil temperature to keep well above air temperature and lichens may be active under layers as much as 30 cm deep (Kappen in Friedmann 1993). On the other hand, it retards heating up in spring. Late-lying snow provides moisture for summer growth but if it persists too long it reduces the growing season. The effects of lying snow on plants are consequently highly variable, depending both on the length of the summer and the preferences of individual species.

Topography is crucially important. Protection from prevailing winds and an orientation which gives optimum exposure to solar radiation makes all the difference. Because of such effects of small-scale topography, areas in which bryophytes, lichens, or algae find moisture and warmth sufficient for growth in close stands are to be found alongside what may properly be called desert. Once established, the plant cover provides insulation and promotes the survival and development of the community.

4.3.2.1 Arctic cold-polar desert

This occurs mainly as a fringe around the Arctic basin with its southern limit at the 2 °C isotherm. Cape Chelyuskin (77°44′N 103°55′E) on the Taymyr Peninsula is in this zone, having a mean air temperature above zero only in July and August, reaching 1.8 and 0.6 °C, respectively. Nevertheless, microclimate temperatures may get considerably higher. In northern Greenland at 82°29′N in May, 3 °C was recorded within a tuft of saxifrage and 10 °C within a moss cushion when the air temperature was −12 °C at noon. During winter the snow cover is often sparse. The soils are of the protoranker type with desert pavement on the bare surfaces between patches of vegetation. While some soils are neutral, others are saline or alkaline. Mosses or lichens predominate but cover as little as 2% of the surface. Various species of crustose and foliose lichens occur on rocks and where rock texture is suitable there are endolithic communities of lichens and algae. There is a scattering of flowering plants, including the Arctic poppy, moss campion (*Silene acaulis*), *Draba* spp., and *Saxifraga* spp. (Sage 1986; Pielou 1994; Crawford 1995; Figs 4.10, 4.11). They form dense tufts or flattened cushions which may be as much as a hundred years old and have two to six times as much dry weight in their underground parts, which store carbohydrates, as in stems and leaves. Polar desert soils contain a considerable seed bank and can produce a surprising display of flowers under favourable conditions.

The species variety of the microfauna varies with the moisture content of the soil, populations often being highly localized. They include protozoa, tardigrades, nematodes, enchytraeid worms, mites, and collembola (springtails). These invertebrates feed on bacteria, algae, fungal hyphae, and debris, but not on living plant material to any extent. Vertebrate herbivores such as the Arctic lemming (*Dicrostonyx groenlandicus*), which feeds on mosses and lichens amongst other things, and the Arctic hare, the main food of which is Arctic willow (*Salix arctica*), may stray into polar desert but it is not their favoured habitat. Likewise, the carnivorous snowy owl (*Nyctea scandiaca*), which has been seen at 82 °N in winter, is only an occasional visitor.

Fig. 4.10 Dry Arctic tundra; a mosaic of foliose and fruticose lichens with cushions of moss and flowering plants, Magdalenafjord, Spitsbergen, *c.* 79°40′N 11°30′E. (Photo G. E. Fogg.)

Fig. 4.11 Arctic poppy in dry Arctic tundra, Longyearbyen, Spitsbergen, 78°12′N 15°40′E. (Photo G. E. Fogg.)

It is difficult to estimate either biomass or primary productivity in such a patchy and heterogenous community. A common method is to harvest the above-ground biomass and determine its dry weight but it is desirable to distinguish between living and dead material and to take account of plant biomass underground. Knowledge of the changes in morphology with age or assumptions about the relation of photosynthesis to standing crop, may be used to obtain estimates of primary production. Together with meteorological data, short-term laboratory or *in situ* determinations of radiocarbon-labelled carbon dioxide uptake or infrared measurements of carbon dioxide concentration changes may be used to calculate primary production on a daily or annual basis. It is felt, albeit difficult to prove, that determinations of assimilation rates give more reliable estimates than even the most elaborate of harvesting techniques (Wielgolaski 1975). Representative values of biomass and primary productivity are given in Table 4.2. The live plant biomass is not much less than that in more favourable habitats in the same zone but primary productivity is lower by an order of magnitude, the overall efficiency in utilization of light being only 0.03%. This reflects the paucity of opportunities for photosynthesis and growth in these desert areas.

4.3.2.2 Antarctic cold-polar desert

The dry *fell fields* of the Antarctic are characterized by cushion-forming mosses and patches of lichen with areas of bare ground (Smith 1972). Again, the soil is of the protoranker type. Larger pebbles in the desert pavement, if

Table 4.2 Biomass of vegetation (g dry weight m^{-2}) and rate of net primary production (g dry weight $m^{-2} yr^{-1}$) in terrestrial polar habitats

Vegetation zone and type	Living biomass (standing dead)			Net primary production
Frigid-polar; Antarctic dry valley desert (4)	4.4–7.4[a]	(9.5–214)[b]		0.01
Cold-polar; Arctic desert (2)	125–185			3
Antarctic desert (3)	200			100 (?)
Antarctic moss turf (3)	291–969			321–497
Antarctic moss carpet (3)	156–204			226–548
Cool-polar; Arctic sedge moss meadow, tundra	959–2083	(414–1125)		185–280
sub-Antarctic meadow (1)	937	(1598)	1642[c]	840
sub-Antarctic dwarf shrub (1)	1521	(517)	7536[c]	1605
sub-Antarctic tussock (1)	7525	(5005)	5000[c]	6025
Mild-polar; sub-Arctic moss, lichen, dwarf shrub (2)	1100–2350			100–600
sub-Arctic shrub (2)	1400–5850			500–1000
sub-Arctic mire (1)	892	(450)		153

[a]From lipid phosphate determinations; [b]from organic matter determinations; [c]below ground, living, and dead.
Data from (1) Rosswall and Heal 1975; (2) Wielgolaski 1975; (3) Laws 1984; (4) Nienow and Friedmann in Friedmann 1993.

translucent, may shelter hypolithic communities. Frost heaving occurs and the ground is often patterned. Nevertheless, if the substratum is stable, colonization by lichens can be surprisingly rapid. On Signy Island, lichen dominated fell field can be established in 10–20 years if suitably weathered rock surfaces are available. The lichens, which are often colourful, include crustose *Caloplaca* spp., fruticose *Usnea* spp., and foliose *Umbilicaria* spp. Some specimens are of a size which, assuming growth rates have not altered much over the years, suggest that they are 300–600 years old. The mosses, with sparingly branched erect main shoots radiating in dome-shaped colonies, include species of *Andreaea* and *Polytrichum.*

The southernmost patches of fell field lie under the 'ozone hole' (p. 20). Lichens and, perhaps to a lesser extent, mosses, produce specific substances in response to exposure to UV-B which absorb this radiation and provide some protection against it. There is an inverse relationship between the concentrations of these substances in specimens of *Bryum argenteum* collected in Antarctica and the mean December ozone levels above the South Pole. Experiments carried out in the sub-Arctic show that increased UV-B irradiation has variable effects on plants, stimulating the growth of some species and inhibiting others. Although it accelerates the degradation of dead surface litter, it retards microbial decomposition processes (European Commission 1995, Björn *et al.* in Woodin and Marquiss 1997).

The fauna of dry fell fields is relatively poor; apart from protozoa it includes nematodes, collembola, and acarine mites (Block in Laws 1984). Collembola occur in densities ranging from around 47 to 3330 individuals per m^2, numbers increasing with the moisture content of the soil. *Gomphiocephalus hodgsoni* feeds on detritus, filamentous cyanobacteria, microalgae, and fungal hyphae according to availability but does not take living macroalgal or moss tissue (Davidson and Broady 1996). It is most active between 7 and 15 °C but its lower limit is between −11 and −23 °C. A mite, *Nanorchestes antarcticus*, is reported as the southernmost occurring animal, at 85°32′S 153°W, and also has the widest distribution of any Antarctic arthropod. Densities are in the range 20–2000 individuals per square metre. *Nanorchestes antarcticus* has a remarkably high heat tolerance with an upper lethal temperature of 37.2 °C at relative humidities greater than 90%, but with a lower limit between −23 and −41 °C. The sensitivity of terrestrial arthropods to increased UV-B flux at high latitudes seems to be unknown. Being mobile, they can, of course, take evasive action.

The success of these and, indeed, all polar animals, depends on having a cyclicity adapted to environmental periodicities, particularly in water and temperature. The mite *Alaskozetes antarcticus* (Fig. 2.5; Block in Laws 1984) provides an example of this. Adults can live for over a year and females can produce eggs on several occasions. This results in a mixed population of adults, larvae, and the three nymphal stages at any one time. The risky periods are during the hatching of eggs and the four moults needed to complete development, all of which must take place in the summer. There is much mortality in the winter but the juveniles are at least as hardy as the adults. This strategy allows egg production to occur whenever conditions are favourable with subsequent successful development to adulthood and so enhances the survival potential of the species. A species having a simple annual cycle might well be eliminated by a succession of severe summers.

Adjacent to this polar desert and often mixed with it in a mosaic are various types of community with more closed cover. Smith (1972) has classified many such communities on Signy Island. Of these, the most developed is the Antarctic herb tundra formation, which is characterized by the presence, either separately or together, of the two native flowering plants of the Antarctic, the grass *Deschampsia antarctica* and the dicotyledon *Colobanthus quitensis* (Fig. 4.12). These occur on north-facing slopes on moist soils comparatively rich in nutrients. Sometimes *Deschampsia* forms a sward of several square metres. *Colobanthus* occurs as discrete cushions. Mosses, but few lichens, accompany them. Both species are capable of producing viable seed, although not every year. Vegetative propagation is important. Birds such as dominican gulls (*Larus dominicanus*) and cape pigeons disturb established clumps and uprooted pieces may be distributed by wind or gravity to establish themselves elsewhere.

Fig. 4.12 *Deschampsia antarctica* (above) and *Colobanthus quitensis*, showing previous year's seed capsules (centre), fell field, Signy Island. (Photo G. E. Fogg.)

Investigation of two communities on Signy Island over the period 1969–81 provided information on environmental factors, primary and secondary production, and decomposition processes under cold polar conditions. One reference site was a well-drained moss turf, consisting principally of *Polytrichum alpestre* and *Chorisodontium aciphyllum*, but heavily colonized by lichens. It occupies a favourable northwest slope and the accumulation of acid peat (pH 4.7) below it raises the vegetation clear of most drainage water. In places on Signy Island such peat is 1–2 m deep and the base of one bank has been dated by the radiocarbon technique at around 1840 BP. The progressive colonization by epiphytic lichens and their removal by frost and wind action results in large expanses of loose, eroding, peat. The other reference site, in an adjacent shallow basin and containing a carpet of the prostrate branched mosses *Calliergidium austro-stramineum*, *Calliergon sarmentosum*, and *Drepanocladus uncinatus*, is on ground which receives melt water from snow and is always wet and sometimes waterlogged. Here, the peat thickness is variable, 4–20 cm. A long-term programme of microclimate monitoring was set up on both sites using sufficient areas to ensure that damage during sampling and inspection was minimal. Quadrats were photographed routinely to follow changes in vegetation. Analyses of standing crops and transfers of organic matter within the communities were made using data for biomass, production, and respiration of plants, microflora, and invertebrates. From these, annual rates of consumption, egestion, assimilation and production of the microflora and microfauna were estimated and the efficiencies of transfer between trophic levels calculated. The flow diagram for the moss turf in Fig. 4.13 shows a simple trophic structure, without grazers and few micropredators (Davis 1980, 1981). The rates of primary production (Table 4.2) are within the normal range. Less than 0.04% of it is consumed directly, the food of the invertebrates being algae, other microorganisms, and detritus. The moss carpet has a similar trophic structure but lower primary productivity than the turf and much less activity by mites and collembola. On the other hand, it has a higher rate of microbial decomposition, as is shown by the smaller accumulation of peat. The difference in decomposition rates estimated from oxygen uptake was borne out by observations on the decomposition of buried cotton strips. Mathematical models based on the flow diagrams can be used to predict aerobic decomposition rates. The observed decay rate in the moss turf was higher than predicted, possibly because it included below-ground plant respiration. In the carpet the observed rate was too low and this is almost certainly attributable to anaerobic respiration and decomposition not being taken into account. The peat does sometimes become depleted of oxygen and supports an anaerobic microflora. This includes methanogens, which oxidize low molecular weight organic compounds produced by other anaerobic organisms, using carbon dioxide as a hydrogen acceptor and so reducing it to methane. In *in situ* experiments the moss carpet was found to release an average of 1.24 mg C m^{-2} day^{-1} as methane during the summer.

A parallel study of mineral cycling in the Signy reference sites has not been done but an adequate supply of the major plant nutrients seems to be available. The numerous birds and seals on the coast, less than 0.5 km away, provide supplies of nutrients, particularly nitrogen and phosphorus, which are spread by various agencies. Precipitation brings down volatile nitrogenous compounds emanating from the colonies. The frequent high winds

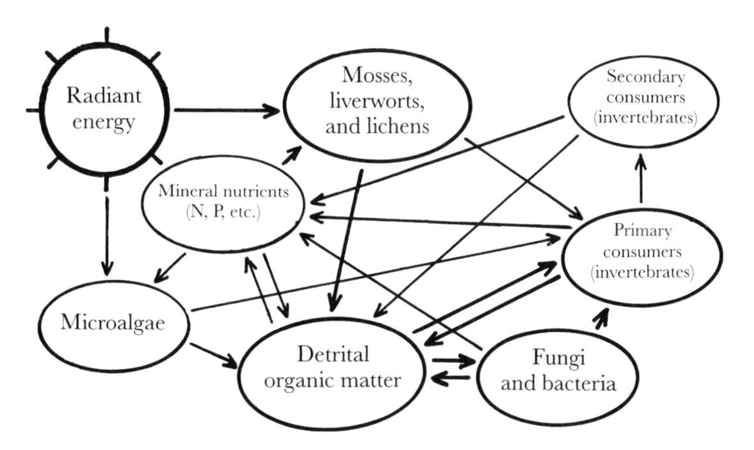

Fig. 4.13 Flows of energy and materials in moss turf on Signy Island. (Based on data from Davis 1980, 1981.)

distribute inorganic and organic particles. Sea spray provides sodium, potassium, magnesium, and chloride. Random droppings from overflying birds and wandering penguins are important sources of nutrients which can be concentrated by build-up in snow before becoming available in run-off. Biological fixation of nitrogen is appreciable in base-rich areas but contributes little in these moss communities. There appears to be little seasonal fluctuation in extractable nutrients.

4.3.3 Cool-polar tundra

Highest mean monthly air temperatures in the cool-polar zone vary between 3 and 7 °C. However, it is longer growing seasons rather than temperature itself which allow greater abundance of plant and animal life than is found in the cold-polar zone. Dramatic changes in microclimate take place with the transition from a snow-covered to a snow-free surface which initiates the growing season. There is an equally abrupt reversal in autumn. The seasonal cycle in energy balance in Arctic tundra near Barrow (71°18′N 156°40′W), shown in Fig. 4.14, is paralleled by one in physical appearance of the tundra surface. There are less dramatic differences related to microrelief and vegetation structure.

4.3.3.1 The physiological ecology of tundra plants

Before considering communities further, more must be said about the effects of polar conditions, as distinct from freezing, on plants. These effects are complex. Surprisingly, the success of a polar plant depends more on tolerance of summer frosts than on withstanding winter freezing. Different processes involved in growth, such as germination, ion uptake, photosynthesis, differentiation, and reproduction have different temperature coefficients and optimum temperatures so that the response of the plant as a whole depends on how well they are orchestrated in harmony with environmental conditions. Most physiological activities of polar plants are less sensitive to low temperature than they are in their temperate counterparts. Photosynthesis, which may be perceptible at −10 °C, usually has an optimum temperature of around 15 °C as compared to 25 °C in warmer climes. This low temperature optimum seems to be related to high RuBP-carboxylase (p. 34) content. This enzyme has the disadvantage of having twofold activity; as well as being a carboxylase (fixing carbon dioxide) it is an oxygenase which brings about extra respiration in light, thereby consuming products of photosynthesis. This light respiration is not significant at low temperatures but when Arctic plants are transplanted south their high RuBP-oxygenase activity makes survival impossible. Chlorophyll contents are not any higher in polar plants than in temperate or tropical species. We would expect this since the function of this pigment is light absorption, which is temperature-independent. The annual amount of carbon assimilation is limited mainly by the length of the growing season and to a less extent by temperature, light

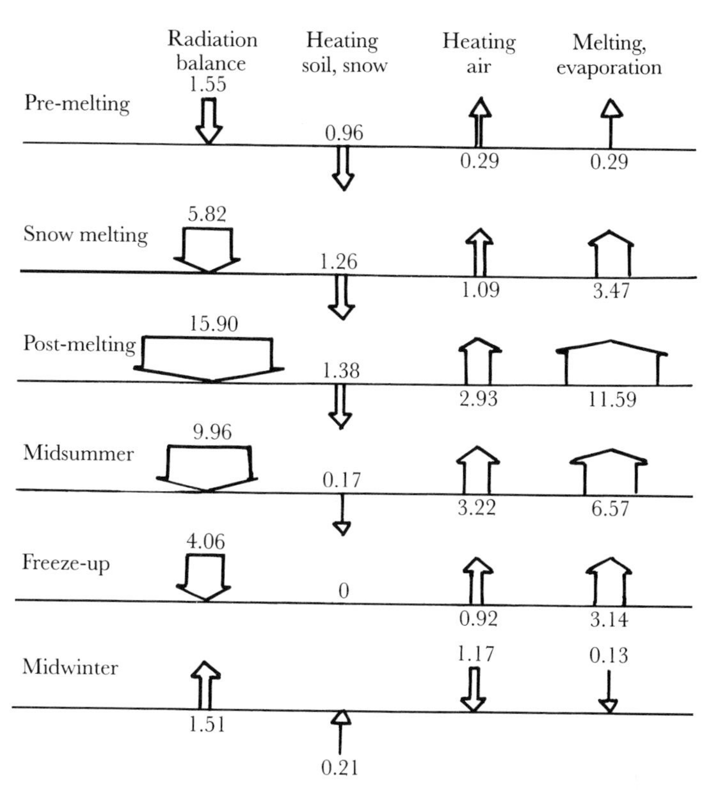

Fig. 4.14 Heat balances for six characteristic periods in Arctic tundra at Barrow, Alaska. The width and direction of the arrows and numbers at the base of each arrow indicate energy flux directions and rates, MJ $m^{-2}day^{-1}$. (Redrawn from Weller and Holmgren 1974, by courtesy of the authors and the editor of *Journal* of *Applied Meteorology*.)

intensity, or nitrogen supply. Nutrient uptake is limited more strongly by availability than by temperature. The *in situ* growth rate of Arctic plants is comparable to, or even higher than, that of similar temperate plants growing in ambient temperatures 15–20 °C higher. Low temperatures may, however, restrict shoot elongation, as is seen if a willow twig is propped up above the warm microclimate close to the ground, when its increase becomes reduced by 30 to 40%. Nevertheless the limiting effects of cold on Arctic plants seem to be exerted more by reducing the rates at which resources become available than by direct effects on growth. Thus, low temperatures maintain permafrost which restricts drainage, resulting in anoxic conditions. Cottongrass, *Eriophorum angustifolium*, is adapted to this situation by having anoxia-tolerant rhizomes which can grow deeply enough to be secure from frost dehydration injury. Anoxia also reduces rate of decomposition and hence nutrient supply (Fig. 4.15; p. 99). Covering a plot with a ventilated plastic cloche (see Fig. 2.1) has dramatic effects in increasing growth but this is difficult to interpret since, apart from raising air

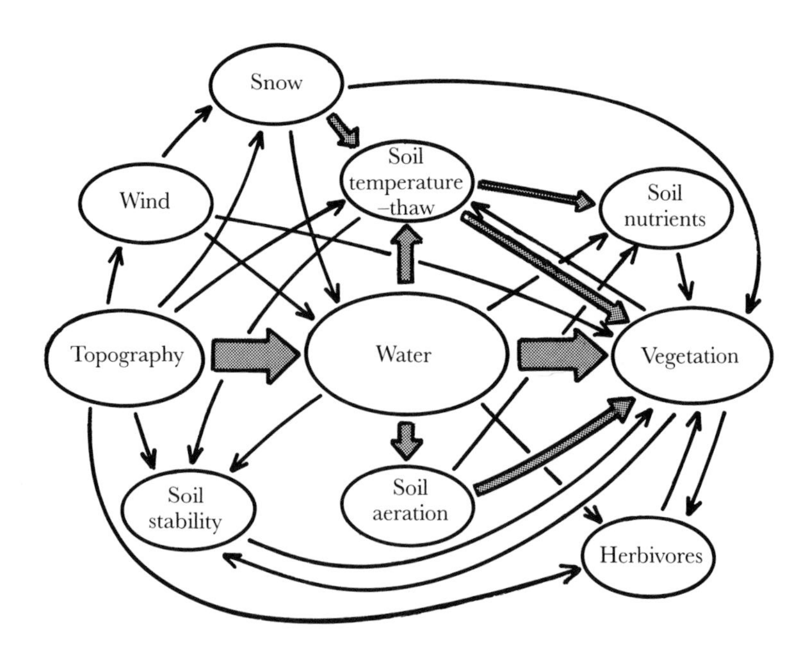

Fig. 4.15 Interactions between the principal environmental factors affecting tundra vegetation. The strength of control is indicated by the thickness of the arrow. (Modified from Webber 1978, see Reynolds *et al.* in Reynolds and Tenhunen 1996.)

and soil temperatures by several degrees, cover affords protection from wind, raises relative humidity, and lengthens the growing season. The importance of these various effects varies with the nature of the vegetation. Similar effects are produced naturally by 'snow hothouses', which are cavities between the lower surface of the snow and the soil, formed as a result of heating of the soil by radiation penetrating the snow. The snow above the cavity crystallizes and turns to ice, covering the minature greenhouse like a window. The temperature may rise 2–5 °C, or sometimes as much as 10 °C, above that of the outside air. These 'hothouses' are characteristic of Arctic tundra in spring and are decisive for the germination and development of some plants by allowing them to start growth when air temperatures are below freezing.

Although it cannot be considered in detail here, it should be noted that the success of an Arctic plant species depends greatly on its plasticity, both phenotypic and genotypic. The purple saxifrage, which grows further north than any other flowering plant, has specialized forms with opposite strategies for survival. These are adapted to either cold, wet, or warmer, drier, microhabitats and are genetically distinct but interfertile, providing a mutual support system for survival (Crawford 1995). Polyploidy – duplication of chromosomes – which rises in frequency the higher the latitude, occurs in as much as 85% of plant species around 80 °N (it also occurs in animals, e.g.

freshwater planktonic crustacea). It is believed to confer increased adaptability to extreme conditions.

Completion of its life cycle by production of viable seed is made difficult for a polar plant by the brevity of the growing season. Annuals must do it in one season and are rare in the Arctic. Biennials, requiring two seasons, are more common but most polar plants are perennials. By drawing on reserves accumulated during the previous season (p. 80), by having wintergreen leaves that can begin photosynthesis without delay in spring, and by producing in late summer flower buds which can overwinter to bloom the next season (p. 32), polar plants contrive to flower except in the most unfavourable summers. There is then the problem of pollination in order to produce seed (Chernov 1985). Many plants, including grasses, sedges, and rushes, are wind-pollinated but over 50% of Arctic species, including some of the commonest, are insect-pollinated. Some have flowers adapted to employ specific kinds, usually bumblebees or butterflies. Most, however, can be pollinated by a variety of 'flies', among which even mosquitoes may figure. In addition to nectar and pollen as food, some Arctic flowers add the attraction of warmth (p. 37). For many species cross-pollination is not essential and self-pollination can result in viable seed. A considerable number of Arctic plants (e.g. grass species of the genus *Poa* and *Potentilla* spp.), goes further and avoids the problem altogether, reproducing by *apomixis*, in which seed is produced without fertilization. With all these it sometimes happens that an unfavourable summer prevents production of viable seed but, since the seed of most Arctic plants can lie dormant in soil for many years, this need not affect the survival of a species in a given habitat (p. 80). As an alternative to sexual and quasi-sexual reproduction, vegetative multiplication by means of rhizomes or the formation of bulbils in place of flowers are important means of survival for many polar plants (p. 83). This has the disadvantage of producing clones of genetically identical plants with sacrifice of the capacity for genotypic adaptation to changing conditions (Sage 1986; Pielou 1994).

4.3.3.2 Arctic tundra

In this the vegetation varies between upland dry heath-like tundra and wet meadow. Wet tundra with continuous plant cover is the most extensive of vegetational types in the Arctic, owing its existence more to permafrost than precipitation. There is sharp distinction between these wet and dry habitats, the intermediate, *mesic*, condition, in which the soil and vegetation have developed to buffer extremes in water supply, being generally absent.

A transect from the crest of a hill down into a valley gives a picture of the general pattern. The results shown in Fig. 4.16 were obtained on the Imnavait Creek watershed (68°36′N 149°20′W) in Alaska in an intensive project to provide basic understanding needed for management of tundra threatened by development of the oil industry (Reynolds and Tenhunen

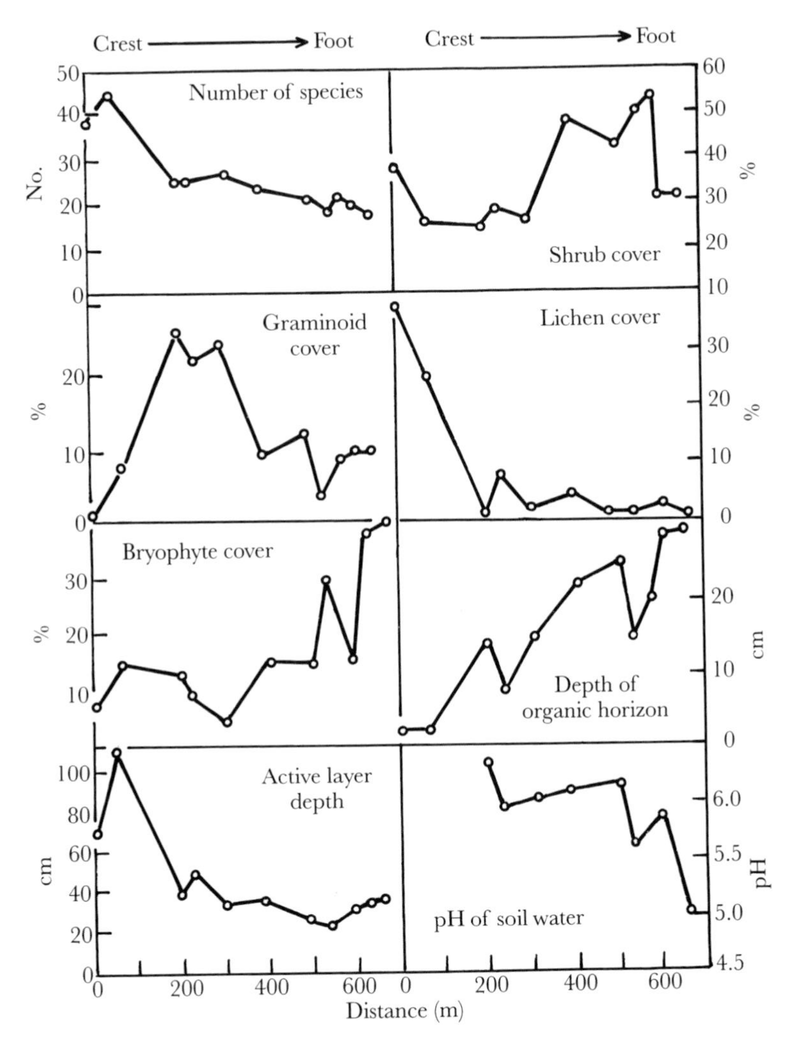

Fig. 4.16 Changes in vegetation types and soil characteristics along a transect from the crest downslope in the Imnavait Creek watershed. (Redrawn from Walker and Walker in Reynolds and Tenhunen 1996, by courtesy of the authors and Springer-Verlag, Berlin.)

1996). The crest fell field has protoranker soil with frost scars and supports a varied flora with lichens predominating and abundant prostrate ericaceous species. Going downhill there is a general decrease in species diversity and a steep fall in lichen cover. The thickness of the organic layer in the soil increases and its pH and concentration of nutrients fall. Shrub cover is less mid-slope, rising to a maximum near the bottom but diminishing again in the mire at the foot. Grasses and sedges are most abundant midway but moss cover, which is fairly uniform over most of the slope, increases steeply near the bottom. This has important effects on the soil environment. The insulation provided keeps the permafrost from thawing and inhibits frost heaving. In spring and summer the moss acts as a sponge and retards drying out. Wherever the bog moss *Sphagnum* appears, as it does here, its great water-holding capacity promotes the formation of mire.

The components of the flora differ but this pattern is much the same throughout the cool-polar Arctic. On Devon Island (75°10′N 85°00′W) the crest zone has a cover consisting of lichens, about 38%, flowering plants such as *Dryas*, *Saxifraga*, *Salix*, and *Carex*, 20%, and mosses 5%, leaving 37% bare ground. *Sphagnum* is not present in the sedge moss meadow but, as at Imnavait Creek, the dominant flowering plants, few of which are annuals, are sedges (*Carex* spp.), cottongrass and dwarf willows (Bliss in Rosswall and Heal 1975). At Marii Pronchishchevoy Bay in Siberia (75°30′N 113°00′E) the fell field has a rich flora of lichens and flowering plants, with prostrate mats of *Dryas octapetala* covering about a quarter of the surface (Matveyeva *et al.* in Roswall and Heal 1975). Sedges are absent from the meadows, being replaced by woodrushes (*Luzula* spp.), and grasses; a dwarf willow (*Salix polaris*) is widespread and *Sphagnum* spp. are again absent.

Microorganisms are abundant in tundra soils. For Devon Island sedge moss meadow, numbers of bacteria determined by plate count – almost certainly giving an underestimate – were around $480 \times 10^6 g^{-1}$ dry weight of soil in the surface 5 cm with a total biomass down to 15 cm of 0.53 g dry weight m^{-2}. The crest of a ridge had 12×10^6 bacteria g^{-1} in the surface 5 cm and a biomass of $0.062 g m^{-2}$ (Bliss in Rosswall and Heal 1975). At Barrow, plate counts of wet meadow soil gave 7.7×10^6 bacteria g^{-1} and direct counts, which include a high proportion of dormant cells, between 3.5 and $10.4 \times 10^9 g^{-1}$, but the biomass was estimated at only 0.002 g dry weight m^{-2} (Bunnell *et al.* in Rosswall and Heal 1975). As to be expected in acidic soils, microfungi are also abundant. On Devon Island their biomass amounted to between 4.5 and $20.3 g m^{-2}$. Sterile forms are the major component and they seem to be cold-tolerant rather than cold-adapted (Bliss in Rosswall and Heal 1975). Other microorganisms recorded from Devon Island are cyanobacteria, at about 1.8×10^5 cells g^{-1}, and protozoa, at 5.3×10^8 cells g^{-1}. Protozoa were estimated to produce 2–6 generations a year, amounting to a total biomass production of $1.5 g m^{-2} yr^{-1}$.

The soil fauna, apart from protozoa, includes nematodes, rotifers, enchytraeid worms, tardigrades, various crustacea, collembola, and diptera, but, notably, not earthworms. Nematodes, producing perhaps 0.16 g dry weight of biomass $m^{-2} yr^{-1}$, are next after protozoa in order of activity in sedge meadow on Devon Island. The total invertebrate production is about $2.6 g m^{-2} yr^{-1}$, consuming bacteria, fungi and decaying organic matter rather than living plant material (Bliss in Rosswall and Heal 1975).

Among the most conspicuous, although they consume only about 0.1% of the above-ground production of plants such as *Dryas* and *Salix*, are the 'woolly bear' caterpillars of the moths *Gynaephora* spp. They persist as larvae for up to 14 years, passing the winters in a frozen state. The adult moths do not feed and are short-lived, being killed by frost. Flying insects, such as these moths and butterflies, avoid being blown away by keeping low in the calmer air, just above the vegetation (p. 68). Diptera are of particular importance in the Barrow area. Three species of cranefly (tipulids) are abundant. The larvae of one of these, the predatory *Pedicia hannai*, sometimes exceeds 200 individuals m^{-2} in wet areas, and a larger species feeding on fungi, *Tipula carinifrons*, reaches over $100\,m^{-2}$ (Bunnell *et al.* in Rosswall and Heal 1975). Where moss cover is continuous, roots of flowering plants branch strongly in a horizontal plane within it and provide a habitat for a variety of invertebrates of different species from those in the soil.

The macrofauna of Devon Island includes one large herbivore, the muskox (*Ovibos moschatus*, Fig. 4.17; Sage 1986; Pielou 1994). This is one of the most specialized of Arctic terrestrial mammals, well adapted to the severe conditions of the polar desert and tundra. Its remarkably thick coat provides insulation sufficient for it to spend much of its time resting rather than expending energy in foraging. In this way, year-long existence in areas of low productivity is possible. In the northeast of Devon Island during 1970–3 the total population varied between 149 and 171. It fed mainly on sedges, removing less than one per cent of the available herbage and having little impact on the vegetation. The Arctic hare generally feeds on *Salix arctica*, taking only the growing tips, and, this plant not being abundant in the area,

Fig. 4.17 Muskoxen. (Photo by courtesy of F. Bruemmer.)

its numbers are low. The Arctic lemming (Sage 1986; Pielou 1994) is present in small numbers in most years. Moss cover favours the construction of their burrows and winter nests are made of sedge roots, mainly in slopes where the snow cover is deepest. Reproduction can occur in winter, as well as in summer, if nests are sufficiently well insulated. Mosses, sedges, and grasses are eaten to a minor extent, the main food being dicotyledons. Snow thaw may be catastrophic for lemmings and other small mammals, the young being drowned in their nests by melt water. On Devon Island snow geese (*Anser caerulescens*) are the only important avian herbivores but their numbers are low (Bliss in Rosswall and Heal 1975). In the Barrow area the principal herbivore is the brown lemming (*Lemmus sibiricus*, Sage 1986; Pielou 1994). Its numbers vary cyclically with a periodicity of 3–5 years, grazing pressure altering correspondingly from less than 0.1% up to 25% of the above-ground net primary production. At a population peak, with a density of perhaps 200 ha^{-1}, there is widespread destruction of the habitat. Their grazing changes the floristic composition of the tundra, encouraging monocotyledons with relatively protected growing points and a capacity for vegetative reproduction. Other vertebrate herbivores have little impact in the Barrow area. Geese, Arctic lemming, and caribou, although common elsewhere on the tundra, are not numerous and hares are absent (Bunnell *et al.* in Rosswall and Heal 1975). The Taymyr Peninsula has one of the largest remaining populations of wild reindeer. The calving grounds are on the tundra but the animals migrate south to overwinter in forest. Reindeer feed extensively on lichens during winter and early spring but turn to wet meadows and dry heath in summer. Their movement being controlled by availability and quality of food supply their impact in any one place is not excessive (Chernov 1985).

The low populations of herbivores on Devon Island limit carnivore populations such as those of Arctic fox (*Alopex lagopus*). The main insectivores are Lapland and snow buntings (*Calcarius lapponicus* and *Plectrophenax nivalis*). Around Barrow, lemmings are preyed on by a variety of carnivores including pomarine skuas or jaegers (*Stercorarius pomarinus*), snowy owls, and Arctic fox. The numbers of these predators decrease when lemming populations are small. In the Taymyr Peninsula, predation by wolves and man contribute, together with natural mortality and immigration–emigration, to fluctuation in numbers of reindeer. Insects are the main food of tundra birds. Amongst shore birds, the sandpipers (*Calidris* spp.) are conspicuous insectivores. They arrive at the beginning of spring melt and start breeding immediately, the young hatching during the mid-season peak in adult insect activity. In the Barrow area, four species of sandpiper consumed in a season some 14 000 cranefly larvae ha^{-1}, about 1% of the total population. During the brief period of emergence of the adult insects about 30% were taken. Buntings, although partially graminivorous, depend largely on insects for food during their breeding seasons. These birds, in turn, fall to predators when the preferred prey of these are in short supply. By serving as supplementary food, insectivorous birds sustain the impact of carnivores on declining lemming populations (Bliss in Rosswall and Heal 1975).

The relationships between these various types of organism within the tundra ecosystem are shown in Fig. 4.18. The primary production upon which everything else depends is an order of magnitude higher in the dry heath than in the cold-polar desert and in the meadows two orders higher. Little of the living plant materials is consumed directly, the main flow of material being through the decomposition of dead plant remains by the combined action of microorganisms and soil invertebrates. The imput of material is greater than the decomposers, limited as they are for most of the time by low temperature, can assimilate, and the surplus accumulates as peat.

Decomposition results in mineralization but mineral nutrients are in short supply in the tundra. This is shown by the more luxuriant growth of plants around old bones and demonstrated more exactly by increases of growth of between 0.5- and 15-fold after treatment with fertilizer. Phosphorus is usually the principal limiting nutrient. It is conserved within living organisms, notably in mosses, which get most of their supply by leaching from living and standing dead material, releasing it only slowly when they themselves die. Some plants (e.g. the heath, *Cassiope tetragona*), take up phosphorus and other nutrients directly from organic matter decomposed by their associated mycorrhizal fungi. The low concentration of soluble phosphorus

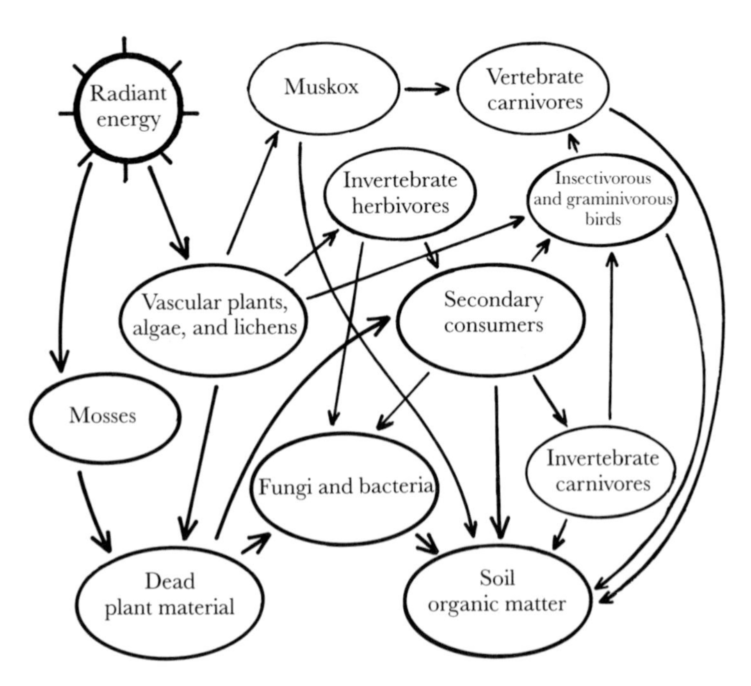

Fig. 4.18 Flows of energy and materials in sedge moss meadow on Devon Island. (Data from Bliss in Rosswall and Heal 1975.)

in the soil results in minimal loss by leaching and run-off. Loss by leaching of combined nitrogen is greater. Fixation of nitrogen by bacteria and cyanobacteria in the soil offsets this to a limited extent, estimated as between 120 and 380 $mg\,N\,m^{-2}\,yr^{-1}$ on Devon Island. Inorganic nitrogen and phosphorus produced by mineralization are quickly released when the soil thaws in the spring but are rapidly taken up and immobilized, mainly in soil microorganisms (Bliss in Rosswall and Heal 1975).

The pattern is altered if conditions favour rodents such as lemmings. Their impact on the living vegetation is sometimes considerable and of the material eaten about 70% is immediately returned to the tundra as faeces. It is not too fanciful to compare the role of lemmings in the tundra with that of earthworms in more temperate regions. Faeces and urine accumulate over the winter and release their contained plant nutrients at the spring melt. The cyclic fluctuations in lemming numbers not only affect plant production and the numbers of their predators but are correlated with nutrient availability, forage quality, and decomposition. Thaw depth is also affected since removal of biomass increases heat transfer into the soil by reducing insulation. There has been a suggestion that in the cycling of nutrients, such as nitrogen, phosphorus, and calcium, these elements are alternately accumulated in lemmings or in plants and soil. Peaks in lemming populations might be supposed to occur following enrichment of vegetation, whereas plant productivity should peak after the maximum number of lemmings has been reached. This nutrient recovery hypothesis is an oversimplification but it is useful as a rough guide.

The interactions of the various populations and processes in the tundra ecosystem are complex but hydrology has an overriding effect. The interrelations of the principal factors controlling vegetation in the Barrow tundra are summarized in Fig. 4.15.

4.3.3.3 Antarctic tundra

The cool-polar lands in the Antarctic are on islands. Compared with the corresponding Arctic zone the oceanic influence makes for cooler summers and milder winters while the lower latitudes, despite extreme cloudiness, provide a longer growing season. Flowering plants assume dominance in grass heath, herbfield, and tussock grassland but the presence or absence of dwarf shrub heath hinges on the fine distinction between a woody shrub and a *suffructicose* herb (e.g. like a wallflower). Ferns and large rosette-forming flowering plants are a significant part of the flora. There are no native land mammals. Again, we have to deal with mosaics of communities but these are rather different from island to island.

Macquarie Island (54°38′S 159°55′W) has a remarkably uniform temperature regime, the monthly mean varying from 3.3 to 7.0 °C over the year. Annual precipitation is 895 mm^3, the mean daily sunshine 2.2 hours, and the mean windspeed 33.5 $km\,h^{-1}$ with maximum gusts up to 128.5 $km\,h^{-1}$.

Snow rarely lies for over a week. Most of the island is a plateau at around 200 m and conditions on this are more severe than at 6 m, the elevation of the meteorological station. There is no permafrost (Selkirk *et al.* 1990).

Short grassland dominated by *Festuca contracta* occurs on coastal terraces but the most striking grassland is the tussock, growing at lower levels on disturbed and well-drained peaty soils. These large grasses, a much as 2 m high with individual plants developing a pedestal of roots and foliage, are a characteristic dominant in the coastal regions of many sub-Antarctic and cool temperate islands in the southern hemisphere. The species differ. On South Georgia it is *Poa flabellata* (Fig. 4.19), on Macquarie *P. foliosa* forms tall tussocks and *P. cookii* and *P. litorosa*, smaller ones. *P. foliosa* forms dense stands with *Stilbocarpa polaris*, a large-leaved plant with a fleshy rhizome which readily regenerates. This community has high biomass and primary productivity (see Table 4.2).

The microbiology and invertebrate fauna of tussock grassland have been little studied. Standing dead material takes several years to decay and humus accumulates in the soil. Earthworms may be present, nematodes are abundant. Tussock provides an attractive habitat for many mammals and birds, affording shelter and a potentially rich source of food in the carbohydrate reserves laid down in shoots and leaf bases. Elephant seals require

Fig. 4.19 Tussock grass, February, South Georgia. (Photo G. E. Fogg.)

none of these but nevertheless find the gulleys between the tussocks comfortable places to lie in. Penguins, petrels, albatrosses, and other birds nest among tussock. These are not herbivores and tussock has flourished on sub-Antarctic islands in the past because there have been no grazing mammals. Where these have been introduced, intentionally as with reindeer on South Georgia and rabbits on Macquarie, or unintentionally as with rats and mice on many islands, they have damaged and, in places, eliminated the tussock. Not only the tussock suffers: on Macquarie, the *Stilbocarpa* in the tussock community is grazed down to stumps by rabbits. The South Georgian pipit (*Anthus antarcticus*) once abundant in the tussock, is now only found where there are no rats. Both rats and mice excavate nests in the tussock stools. The excreta of these various animals produces a habitat rich in nutrients. Growth of tussock on soils unfrequented by seals and birds is noticeably less luxuriant.

The nearest approach to dwarf shrub heath in the sub-Antarctic is the community on South Georgia dominated by the greater burnet (*Acaena magellanica*). South Georgia has a less uniform climate than Macquarie, with mean monthly temperatures at Grytviken ranging between −1.5 and 5.3 °C (see Fig. 1.6). Mean wind velocity is around $15.8\,km\,h^{-1}$ and the mean annual precipitation is 1405 mm. Conditions in the mountains, which go up to 2934 m, are much more severe (Headland 1984).

Acaena is a perennial prostrate herb with stems, woody at the base, arising from long, intertwined, woody rhizomes. It is usually the first colonist on bare screes, being well adapted to stabilize them, and is almost invariably accompanied by *Tortula robusta*, a turf-forming moss, which provides moist, warm, conditions, and humus for the rhizomes and roots. On stabilized ground where a peaty loam has developed, the grass *Festuca erecta* becomes codominant with *Acaena* and on well-drained stable ground replaces it as dominant. On wet seepage slopes the rush *Rostkovia magellanica* accompanies *Acaena*. Biomass and primary productivity in the *Acaena–Festuca–Tortula* heath (Table 4.2) are higher than corresponding estimates from the cold-polar zone but less than those for tussock.

The most abundant invertebrates in the South Georgian communities are mites and collembola, numbered respectively, at around 31 000 and $56\,500\,m^{-2}$. There are also enchytraeids, an earthworm, tardigrades, nematodes, spiders, coleoptera, and diptera. Few are herbivores except for coleoptera, which graze shoot tips of *Acaena* and mosses, and the bases of grass shoots. Spiders and beetles are carnivorous but the larger soil invertebrates are more importantly a food source of overwintering birds on the sub-Antarctic islands (Burger 1985). The activities of bacteria and invertebrates result in rapid decomposition of *Acaena* leaves, 90% loss in dry weight occurring within 14 weeks. Deep peat accumulates only under wet conditions, usually beneath moss. The bog moss *Sphagnum fimbriatum* occurs on only a few sites.

Soils on South Georgia are poor in nutrients except where fertilized by seal or bird colonies. The dead standing material and litter from vascular plants contains as much, or more, nitrogen, phosphorus and magnesium as is in the current year's living leaves. This is released at a rate determined by decomposition and in *Acaena* communities must be rapid. Lichens with cyanobacterial phycobionts may be important in *Tortula-Acaena* stands in fixing nitrogen.

In their natural state the communities of South Georgia are detritus accumulating systems. Rats, introduced in the early 1800s, and reindeer, established between 1909 and 1924, have altered this, especially since confining glaciers have retreated in recent years. Reindeer by grazing and trampling have in places eradicated many of the larger lichens, moss swards, *Acaena*, and tussock. Some overgrazed communities have been replaced by swards of an alien grass, *Poa annua*. Rats eat the roots of tussock, leaves, and rhizomes of *Acaena*, and leaves of some grasses. On Marion Island (46°52′S 37°52′E), mice produce change in soil processes by preying on slugs which bring about the release of nutrients from litter. This increases peat accumulation by stopping the processing of an estimated 1000 kg ha^{-1} of plant litter.

4.3.4 The mild-polar zone

This zone, enclosed by the poleward limit of developed woodland, forms an extensive circumpolar band in the north but is restricted in the south to a small area including the Falkland Islands and a narrow coastal strip of southern Chile. Dwarf shrubs, stems ascending up to 50 cm, as distinct from prostrate woody perennials, such as *Salix arctica* and *Acaena*, are a distinguishing feature of the vegetation. Highest mean monthly air temperatures are between 6 and 12 °C but soils may be frozen for 3–6 months each winter.

4.3.4.1 Arctic mild-polar mire

Stordalen, near the Swedish research station at Abisko (68°22′N 19°03′E) at an altitude of 351 m, is a treeless mire with a microrelief of low hummocks and small shallow depressions (Rosswall *et al.* in Rosswall and Heal 1975). It is *ombrotrophic*, that is, receiving the bulk of its mineral nutrients via precipitation (about 300 mm per year), and therefore *oligotrophic* (i.e. poor in nutrients). Permafrost is near the surface throughout the growing season, from mid-May until mid-September. Snow cover persists for some seven months. The soil is peat, up to 3 m in depth, formation of which began about 5000 BP.

The character of the vegetation is given by the creeping habit of dwarf shrubs such as dwarf birch (*Betula nana*) and crowberry (*Empetrum hermaphroditum*). *Rubus chamaemorus*, the cloudberry, and tussocks of cottongrass accompany these. However most of the cover, some 87%, and much of the above-ground biomass, 61%, is contributed by the mosses *Dicranum elongatum* and *Sphagnum* spp. Low values of biomass and primary production (Table 4.2) as compared with other mild-polar communities reflect the oligotrophy of the habitat.

Insects, both sap-sucking and browsing, and nematodes feed on the plants but only occasionally consume significant amounts. Experiments with cloches in Svalbard show the response of above-ground insects, the aphid *Acyrthosiphon svalbardicum*, for example, increases rapidly with rise in temperature. In contrast, the soil-dwelling invertebrates inside and outside cloches show no significant difference in the short term – perhaps because there is a lowering of moisture content below the cloche. Vertebrate herbivores in Stordalen include reindeer, hare (*Lepus timidus*), Norwegian lemmings (*Lemmus lemmus*), and voles (*Clethrionomys* spp.). Reindeer cross the area during migration but probably do more damage by trampling than by grazing. The small rodents have considerable effects at their population peaks but their consumption is not otherwise important. Grazing is selective, current shoots and opening buds being preferred in early summer, persistent fruits (e.g. of *Empetrum*), in winter. Hares and willow grouse (*Lagopus lagopus*) are heavy grazers. Among predators on these herbivores, bear (*Ursus arctos*), wolf (*Canis lupus*), and wolverine (*Gulo gulo*), have been drastically reduced in number, directly by hunting and indirectly by man's modification of the environment. Numbers of other predators depend on the cycles in rodent populations, for example, red fox (*Vulpes vulpes*) and short-eared owl (*Asio flammeus*), or on massive swarms of insects, for example, willow warbler (*Phylloscopus trochilus*) and meadow pipit (*Anthus pratensis*). Mosquitoes, culicidae, are a nuisance around Abisko in the summer.

Peat deposits are found world-wide, covering about 3% of the earth's land surface and retaining perhaps 460×10^9 tonnes of carbon, more than half of the amount in the atmosphere. Because cold and waterlogging slow down decomposition, most of this peat, about 370×10^9 tonnes of carbon, occurs in the regions dealt with in this book. The mires that produce peat may be ombrotrophic or *minerotrophic* (receiving some part of their water after it has flowed over mineral soil). Sub-polar mires are mostly moderately minerotrophic and support a somewhat greater variety of species than the oligotrophic ombrotrophic kind. Decomposition does not cease entirely within the peat mass; the rate is low but there comes a point when the decomposition throughout the depth is sufficient to balance the addition of matter at the surface, and the deposit stops growing. Most of the carbon that leaves the peat is in methane, which at 5 m depth may reach a concentration of about 10% by volume. Much of this escapes to the atmosphere by mass flow of bubbles. Since both methane and carbon dioxide are 'greenhouse' gases (p. 5), knowledge of sub-polar mires is of importance for the understanding and prediction of global warming. Rise in temperature increases carbon dioxide loss from the soil and so does fall in the water table, since it makes oxygen more available to roots and microorganisms. Warmer and drier conditions thus change tundra from a net absorber of carbon dioxide to a net source, providing a positive feedback for global warming. The effects on methane flux are less certain. Change to warm and wet conditions, giving a deeper active layer and optimum conditions for anaerobic growth, would probably increast methane emission and again provide positive feedback for the greenhouse effect. Drier conditions might cause emission to cease or even turn tundra into a sink for methane (Christensen 1991; Clymo and Pearce 1995; Whalen *et al.* in Reynolds and Tenhunen 1996, Oechel *et al.* in Woodin and Marquiss 1997).

4.3.5 Intrazonal communities

The communities described so far fall roughly into a series corresponding to climatic zones and soil types, extending from extreme to moderate polar conditions. However, there are other communities which, being produced by some locally operating factor, are not distributed zonally. Among these are communities on evaporite soils, on biogenic soils, and in fumarolic areas.

4.3.5.1 Communities on evaporite soils

Polar deserts tend to be saline but soils formed in wet hollows where salts are concentrated to particularly high levels (p. 125) need special consideration. These evaporite soils are usually associated with saline lakes and the salts may come from wind-blown sea spray or by leaching from weathering rocks. Accumulations of salt vary from 0.1 $kg\,m^{-2}$ up to 100 $kg\,m^{-2}$ in the oldest and driest soils. The kind of salt varies with origin but chlorides, nitrates, and sulphates of sodium, potassium, calcium, and magnesium are most common (Campbell and Claridge 1987). Such salinity is inimical to colonization and together with low temperatures may be supposed sufficient to prevent the growth of most organisms. To casual inspection evaporite soils do, indeed, appear sterile. More precisely, numbers of psychrophilic yeast cells in Antarctic dry valley soils decrease with increasing salinity, sites with less than 19.4 μ cation equivalents g^{-1} always contain these yeasts whereas only half those with 19.4–60.5 μ equivalents and no sites with more than 60.5 μ equivalents g^{-1} do. The bacterium *Planococcus*, isolated from a dry valley soil is able to grow in 2.0 M sodium chloride (water potential deficit around 85 bar, p. 41) at a temperature of 0°C. It accumulates amino acids as compatible solutes (p. 36). *Planococcus* has no specific requirement for sodium chloride and its growth rate decreases with increasing concentrations of this salt (Vishniac in Friedmann 1993). A variety of other bacteria have also been isolated from these soils. It may be noted that one biological advantage of high salinity is that, by lowering freezing point, it extends the period of availability of liquid water and reduces the damaging effects of freeze–thaw cycles.

4.3.5.2 Communities on ornithogenic and seal-wallow soils

The habit of the larger polar animals of congregating to breed results in substantial inputs of organic matter on to some coastal areas and inland nunataks. The soils which result have nutrient concentrations which may be so high as to be toxic (Table 4.1). In bird colonies organic matter comes as guano, feathers, and dead birds. In penguin rookeries the deposit generally consists of 10–15 cm depth of reddish-brown material grading into darker organic horizons below and then into mineral matter (Fig. 4.20). The organic content is initially around 30% but the uric acid from guano is rapidly degraded into ammonia, which may reach a concentration of 140 ppm, and the phosphate content is several hundred times that of mineral soils in the neighbourhood. Elephant seal activities produce even higher organic and ammonium nitrogen contents (Table 4.1; Campbell and Claridge 1987; Vishniac in Friedmann 1993).

Given the mechanical disturbance caused by penguins and seals and the anaerobic, toxic, nature of their excrement, no bryophyte or angiosperm is likely to survive in intensively colonized areas unless, as with tussock, the plants are massive and well established. There is an abundance of nutriment for microorganisms and numbers of bacteria may be of the order of $2 \times 10^{10} g^{-1}$, although less than 8% may be metabolically active. Carbon dioxide production by the soil is, in fact, low – about the same as in temperate peat bogs – except when it is warm and moist. The annual loss of carbon is less than 0.5% of the total. Loss of nitrogen from ornithogenic soils is mainly by

Fig. 4.20 The making of an ornithogenic soil. Adélie penguins and chicks, Signy Island, January. (Photo G. E. Fogg.)

volatilization of ammonia. Of the 275 kg of ammonium nitrogen introduced each day into a penguin rookery on Marion Island, some 220 kg is volatilized. Wind and rain are the main agents removing solid material.

Bird and seal colonies benefit adjacent vegetation. An example from the frigid zone of the Antarctic is Robertskollen, a group of inland nunataks at 71°28′S 3°15′E, which have breeding colonies of snow petrels (*Pagodroma nivea*). Although water is the major limiting factor, dispersion patterns show that vegetation cover, mostly of lichens with a few mosses, is about doubled on sites adjacent to the bird colonies, as compared with that in the immediate vicinity of nests, and about seven times as much as that on nunataks without birds. The green alga *Prasiola*, which has high resistance to organic matter, does best on wet patches within the colony. The only free-living arthropods are mites, one species *Maudheimia wilsoni*, being most abundant where plant cover is greatest, whereas another, *Nanorchestes antarcticus*, is more abundant within bird colonies (Ryan and Watkins 1989). On Marion Island, in the cool-polar Antarctic, sites used by elephant seals when moulting show seasonally alternating degeneration and regeneration. The seals deposit faeces, urine, and moulted skin, which result in high nutrient levels favouring tussock grass. On the other hand, a plant sensitive to high nutrient levels, the halophyte *Tillea moschata*, occurs on deserted moulting sites. The most numerous invertebrates are collembola and mites but during the peak moult haul-out their abundance is decreased by trampling.

Ornithogenic and seal-wallow soils are less evident in the Arctic. Presumably because of mammalian predators, birds tend to nest in densities of a few pairs per hectare or in dense colonies on cliffs (Fig. 4.21) and seals to breed on sea ice. Consequently, there is less input of plant nutrients to the land from the sea.

4.3.5.3 Communities in geothermal areas

Fumaroles occur in some places in the Arctic, such as the shores of Bockfjord, Svalbard (79°32′N 13°20′E), and near Mounts Erebus (77°40′S 167°20′E) and Melbourne (74°21′S 164°42′E) on the Antarctic continent, and on the South Shetland (62°57′S 60°38′W) and South Sandwich (*c.* 58°S 25°W) islands. These maintain warm conditions which promote intense chemical weathering to produce clay minerals. Besides warm ground there may be a fairly constant supply of liquid water from steam condensation. In Iceland the effect on growth of vegetation characteristic of mild-polar conditions is clear around alkaline springs, where dense stands of dwarf willows, grasses, herbs, and heaths, develop. The hot springs of Greenland support some species which are unrecorded elsewhere in the region and others which reach their northern limits in the favourable temperatures which are available.

Habitats heated by volcanism in parts of Antarctica present an interesting biological situation since they are isolated by long distances over sea or ice

Fig. 4.21 Colony of little auks, Hornsund Fjord, Spitsbergen (*c.* 77°N 16°E). Note the stripes of more luxuriant vegetation below areas with dense concentrations of nests. (Photo G. E. Fogg.)

from potential sources of colonizing organisms. Nevertheless, on Deception Island in the cold-polar zone in the South Shetlands, two years after an eruption which covered much of the island deep in ash, warmed surfaces became colonized. Mosses and six species of protozoa were among the first colonist. Ten years later the widely dispersed Antarctic testate amoeba, *Corythion dubium*, made its appearance (Smith 1984). Eleven years after the latest of the eruptions these areas supported closed stands of bryophytes covering several square metres which included, in addition to those found in local non-thermal habitats, seven species associated only with fumaroles. In the frigid Antarctic, at an altitude of 2733 m on Mount Melbourne, fumarolic ground supports growths of bryophytes and algae. The last volcanic eruption there probably took place no more than a hundred years ago. The soil varies in temperature between 14 and 31 °C and has a regular moisture supply from condensed steam. It remains free of ice, at least during the summer, but after blizzards may become coated with a crust of ice raised a few centimetres above the surface. Six species of cyanobacteria, of which the most abundant is the nitrogen-fixer, *Mastigocladus laminosus*, and five unicellular green algal species have been found. There were also the moss *Campylopus pyriformis* and the liverwort *Cephaloziella exiliflora*. Heterotrophic microorganisms included thermotolerant fungi, actinomycetes, and thermotolerant and thermophilic bacteria. The only animal found was *Corythion dubium*. The point of interest here is the presence of *C. pyriformis*. Mosses of this genus are primary colonists of volcanic substrata and another species occurs in fumarolic ground on Deception Island but *C. pyriformis* was a new record for Antarctica although it has a circumpolar distribution in south temperate regions. Mount Melbourne is a long way from sources of its propagules and also well south of the westerly air stream that might distribute them (Broady in Friedmann 1993).

4.4 Comparison of Arctic and Antarctic terrestrial habitats and communities

The general patterns are similar in the two regions. In the mild-polar zones, dwarf shrub and shrub communities have annual primary production rates (see Table 4.2) which are not greatly different from those of birch forest in England but only about a tenth of that of tropical rain forest. Proceeding polewards one finds continuous stands of low-growing vegetation becoming patchier with a mosaic of communities related to local geology or topography. In the cold-polar zone, sedge moss meadow on Devon Island is about as efficient (0.79%) in utilization of photosynthetically available radiation as temperate grassland. On the basis both of unit area of ground and of plant cover, biomass and primary production diminish to about the same levels in Arctic and Antarctic as one goes polewards. Liquid water rather than low temperature is the major limiting factor and the moister Antarctic tundra communities exceed those in the Arctic both in biomass and primary production. The deserts of the cold- and frigid-polar zones have extremely low biomasses and primary productivities, photosynthetic efficiency being around 0.03% as compared with 0.4% for land vegetation in general.

Both in the Arctic and Antarctic, lichens and simple algae among phototrophic organisms and mites and collembola among invertebrates are the most hardy. Endolithic communities of phototrophic and heterotrophic microorganisms, largely self-contained, are the forms of life which can achieve some activity and growth under the severest terrestrial conditions which either pole can inflict.

That being said, it is striking what differences there are between Arctic and Antarctic terrestrial communities. This is evident both in species diversity and in the species themselves. An overall view of the data leaves no doubt about this. Thus, against about 900 species of vascular plants in the Arctic there are only 2 in Antarctica and whereas the Arctic has 48 species of native land mammals, Antarctica has none. Nevertheless, it is far from straightforward to make any valid detailed numerical comparisons. It is difficult to match communities north and south, inventories of species are incomplete with gaps in different groups for different localities and one suspects that the expertise and assiduity of field workers has varied considerably. In Table 4.3 an attempt is made to compare data from roughly similar habitats from Arctic and Antarctic cool-polar zones. These two areas have had about the same time since the peak of the last ice age to

Table 4.3 Species numbers in cool-polar zones. Arctic data from Bliss and Antarctic from Smith and Walton (both in Rosswall and Heal 1975)

	Devon Island (75°33′N)		South Georgia (54°55′S)	
	Sedge meadow	Cushion plant, lichen heath	Mesic meadow	Fell field
Grasses, rushes, and sedges	>4	–	1	3
Dwarf shrubs	1	1	1	2
Total flowering plants	29	–	3	6
Pteridophytes	0	0	0	1
Bryophytes	30	10	10	22
Lichens	1	15	0	34

become colonized and it is evident that it is the isolation of Antarctica amid a huge expanse of ocean, whereas the Arctic includes parts of continental land masses, that accounts for the differences in vascular plant and terrestrial mammal species numbers. As Table 4.3 shows, the position is less clear with pteridophytes, bryophytes, and lichens. However, it is significant that these are groups having small wind-dispersable propagules. The two Antarctic vascular plant species do not occur in the Arctic but among the bryophytes and lichens there seem to be a few cosmopolitan species and a significant bipolar element, especially with lichens (Longton 1988). Assuming that there really is genetic identity and apart from the possibility that bipolarity may have arisen by evolution of the same species in two different places, which is not altogether unlikely with lichens, two explanations are possible; long-distance transport of propagules, perhaps hopping via mountain tops across the tropics, or survival from times when Gondwanaland was still linked to land masses which are now in the north.

Attempts to grow Arctic plants from seed under natural conditions in Antarctica have met with little success and, conversely, tussock grass is difficult to establish in the northern hemisphere. This rather surprising state of affairs arises from the great difference between the cold, oceanic, climate of the sub-Antarctic and Antarctic and the continental/sub-alpine climates of the Arctic. An illustration of this difference is seen in the latitudes at which comparable vegetation types occur, those in the south being 20 or even 30 degrees further away from the pole. This means that photoperiods are different. There are also effects on the altitudinal range of plants. At 75°N on Devon Island vascular plants occur on the plateau at 400 m above sea level, whereas on Signy Island, at 60 °S, vascular plants do not occur above 100 m. The nutrient levels in soils are similar in both regions, except that seals and seabirds have a much greater effect in enriching soils and sea spray in modifying them, in the southern hemisphere.

A comparative study of the ecology and primary production of Arctic and Antarctic plants, carried out during the International Biological Programme's bipolar botanical project in 1967–73, found that the short, but favourable, summers at Disko (69°15′N) on the West Greenland coast, restrict development and reproduction whereas the longer, cool, and wet growing season on South Georgia (54°17′S) promotes vegetative production. The Arctic climate has selected species capable of short periods of rapid growth and opportunistic reproduction or rapid reproductive development. On the other hand, the South Georgian climate has favoured species with slow but consistent rates of reproduction and growth. These adaptations have a genetic basis.

Food webs in terrestrial polar regions are of a similar general pattern with the main pathway of energy flow being solar radiation to plants to decomposers to organic matter stored in the soil. Herbivores and carnivores mediate a small flow of biomass and energy in the Arctic but this component is negligible in the Antarctic. The fewer species involved in the Antarctic make for greater simplicity and suitability for meaningful numerical modelling of energy flow systems (Block 1994). Few satisfactory models have been constructed for Arctic ecosystems. However, when one considers matters on the global scale, the vast extent of tundra in the Arctic – some 7×10^6, as compared with less than $5 \times 10^4\,km^2$ in the Antarctic – has more significance. Its peat deposits contain about 11% of the total organic carbon reserves of the world.

In the absence of large terrestrial carnivores, Antarctica provides a habitat for a distinctive fauna, quite different from that of the Arctic. Seven species of penguin, all flightless and nesting ashore in large colonies extremely vulnerable to land-based predators, do not and could not exist in the Arctic. Some Antarctic seals breed on sea ice but three species breed ashore whereas none do in the Arctic. The repercussions of this on the flora of the maritime Antarctic have already been mentioned.

Finally, there is the major difference that the extensive area of moderately productive wetlands in the Arctic provides for large and diverse populations of migrant animals whereas there is scarcely anything of the sort in the Antarctic. Only 11 species of birds are winter residents in the Arctic; the others, around 90 species, must make long migratory flights to other areas where conditions are milder and food is available. Enormous numbers of these migrants assemble in traditional breeding grounds in the Arctic each spring, where, given a good season, ample amounts of plant or insect food are available. Birds, such as geese, arrive with energy reserves, the amount of which determines their reproductive success. The food found in the Arctic must suffice to replenish fat reserves for the return journey. Thus, the sanderling (*Calidris alba*), which migrates from the high Arctic to winter on temperate shores, departs weighing about 110 g as against its normal 50 g. The volume of these migrations is impressive; at Point Barrow (71°20′N

156°40′W), which lies on a favoured migration route, the passage of some 240 000 birds belonging to 50 species was recorded during one 6-week period. The winter destinations are circumpolar, some at great distance. The Arctic tern (*Sterna paradisaea*), for example, is an uncommon but perhaps often overlooked visitor to the Antarctic Peninsula and Baird's sandpiper (*Calidris bairdii*) makes a journey over 100° of latitude from the high Arctic to the Andes of South America (Sage 1986; Pielou 1994).

The reindeer or caribou is one of the most numerous large wild mammals now existing, with a total population of over 2 million. In addition to the feral herds there is more than an equal number of domesticated animals. Apart from those in the northeastern Canadian Arctic, which overwinter on the tundra, individual herds carry out migrations, focusing on restricted calving areas, which may involve twice-yearly movements of many hundreds of kilometres. These movements are partly controlled by availability of food and partly by the need to escape mosquitoes in summer (Sage 1986; Pielou 1994).

Along the Eurasian Arctic coast there are peoples – Lapps, Samoyeds, Yakuts, and others – who migrate, following the reindeer into the tundra in summer and retreating south into forest in winter. This way of life developed in the last two thousand years and has not extended to Arctic America, where the Inuit remained hunters and gatherers until recently. Part of the reason for this is that the Eurasian Arctic is a thin strip across which there is relatively easy passage, whereas the American Arctic presents longer distances interrupted by straits, severely hampering a nomadic lifestyle. Additionally, herding had a long tradition, going back to neolithic times, in the Eurasian Arctic whereas, on the large scale, it was unknown in the American Arctic.

5 The inland waters

5.1 Introduction

There is great variety in the inland waters of the polar regions (Hobbie 1984). It was convenient to consider those contained in ice in Chapter 3, but there is no sharp biological distinction between these and water bodies on exposed land surfaces. Most of the pools and streams which appear on ice are ephemeral but some proglacial lakes, such as Lake Untersee (p. 56), may persist and perhaps eventually become contained in a rock basin beneath. Many lakes in the Arctic and Antarctic may have originated in this way (Priddle and Heywood 1980).

The character of any lake is largely determined by its heat budget. In polar regions an equivalent of the energy given out and dissipated when ice forms in winter must be reabsorbed to melt it, consequently water temperatures have small opportunity to rise much above zero in the brief summers. Ice protects underlying water from wind so that the water column can stabilize and mixing is restricted to the ice-free period. On this basis, polar lakes have been defined as having surface temperatures always below 4 °C, a short ice-free period, and circulation only at the height of summer. Sub-polar lakes have surface temperatures above 4 °C only for a short time in summer and mixing periods, typically, in early summer and early autumn. At midsummer, a thermal gradient may be established, stabilizing the water column, but the *thermocline* between the warmer and colder waters is poorly developed and temporary cooling in summer permits fairly frequent mixing (Hutchinson 1957). As always with such definitions, there are exceptions; the large Icelandic lakes, Thingvallavatn (64°10′N) and Mývatn (65°35′N) are both fed by thermal springs and have water temperatures rising to 10 °C or over for a few weeks in the summer although they are in a sub-Arctic region and have average days of ice cover per year of 95 and 189, respectively.

The differences in distribution of land in Arctic and Antarctic result in rather different types of lakes and different floras and faunas inhabiting them. The difference in flowing waters is extreme. Whereas Antarctica has only a few minor streams, the Arctic lands have some of the world's major rivers flowing through them.

5.2 Arctic lakes

Lakes are abundant in the Arctic but irregularly distributed, most being on the flat coastal plains. Of the many agencies which produce lakes, glacier ice and permafrost have been most active. The excavation of basins in the bedrock and blocking of drainage systems by glacial deposits have produced a large number of small and shallow lakes in northern North America but one, Great Slave Lake (*c.* 61°N 114°W), has an area of 30 000 km^2 and a maximum depth of 614 m. Other large lakes of this type are Nettilling (66°30′N 70°30′W, 5525 km^2) and Amadjicak (65°00′N 71°00′W, 3105 km^2) on Baffin Island. In the mountainous areas valley glaciers have produced lakes of various sorts, including some dammed by ice. These are prone to sudden draining in the warm season, either by an overflow cutting a gorge through the ice or, if the water deepens, by the main body of ice floating up and allowing subglacial drainage.

5.2.1 Glacial lakes

In his account of Char Lake, at 74°43′N 94°59′W on Cornwallis Island, Rigler (1978) made the point that, given adequate resources, including power-driven augers to get through the ice and a warm, well-equipped, laboratory nearby, Arctic limnology is no more difficult, albeit more expensive, than temperate limnology. This lake is 52.5 hectares in area with a mean depth of 10.4 m and a maximum of 27.5 m. The surface is only ice-free in August. During the winter the water column is stratified but is not entirely without circulation because of the sinking of cold, mineral-rich water produced by freezing at the surface. Destratification by solar heating begins in May, although the ice is then at its thickest. In summer the water is isothermal and circulates freely.

Char Lake is extremely oligotrophic, with phosphate evidently the nutrient in limiting supply. The mean chlorophyll *a* concentration in the summer, 0.4 mg m^{-3}, which is among the lowest recorded (Table 5.1), is as predicted from the phosphate concentration. Growth of phytoplankton, as is usual in lakes that are ice-covered for much of the year, is already considerable under the ice in the early summer and peaks soon after the melt takes place. Primary production is much lower than would be expected, the daily production/biomass ratio being 0.05–0.07, whereas in temperate waters it is about unity. This is probably because the phytoplankton is shade-adapted and cannot use high irradiation efficiently (p. 46). The ice is snow-covered until late June, reducing the light penetrating into the water considerably (Fig. 5.1), so that the maximum rate of photosynthesis is reached at about 56 μmol quanta $m^{-2} s^{-1}$, or around $\frac{1}{36}$th of full sunlight. Above this level, photosynthesis is inhibited. The average annual primary production by the plankton is 4.1 g C m^{-2}, of the same order as values for two other oligotrophic Arctic lakes, 0.9 g C $m^{-2} yr^{-1}$ for Lake Peters and 6.6 to

Table 5.1 Chlorophyll *a* concentrations (mg m^{-3}) and primary productivities (g C m^{-2} yr^{-1}) in polar lakes. It should be noted that authors have not always specified whether their estimates are net or gross but the difference is not crucial in this context

Lake, type, and location	Maximum depth (m)	Chlorophyll concentration	Annual primary production	
			Phytoplankton	Phytobenthos
Arctic				
Char, oligotrophic 74°43′N 94°59′W (2)	27.5	0.4	4.1	16
Meretta, eutrophic 74°43′N 94°59′W (1)	9	8	11.3	22.5
Thingvallavatn, oligotrophic 64°10′N 21°10′W (4)	114	0.5–4	95	55–118
Mývatn, S. Basin, eutrophic 65°35′N 17°0′W (3)	4.2	2.5–30	118	272
Antarctic				
Vanda, oligotrophic 77°32′S 161°33′E (6,7)	68	<0.1	0.34	2.5
Bonney, oligotrophic 77°43′S 162°20′E (7)	35	–	78.7	0–113
Moss, oligotrophic 60°43°S 45°38′W (5)	10.5	0.5–8	14.3	6.3
Heywood, mesotrophic 60°43′S 45°38′W (5)	6	2.5–35	173	14.5
Algal, mesotrophic 77°38′S 166°24′E (5)	1	2–9	6–7	172–327
Skua, eutrophic 77°38′S 166°24′E (5)	0.8	5–80	45–133	140–230

Data from (1) Kalff and Welch 1974; (2) Rigler 1978; (3,4) Jónasson 1979, 1992; (5) Heywood in Laws 1984; (6) Vincent 1987; (7) Green and Friedmann 1993.

7.5 g C m^{-2} yr^{-1} for Lake Schrader, both situated around 69°N 145°W at an altitude of 850 m. The main primary producer in Char Lake is not plankton but benthic vegetation, which contributes about 80% of the total. Benthic algae in the rocky shore and deep silty zones are most active, while the luxuriant-looking mosses, as much as 40 cm in length, which grow in beds at depths of between 3 and 15 m, are only about half as productive. The paucity of plankton means that the benthos is able to get adequate light at depths below the reach of winter ice but prolific growth of epiphytes has a shading effect. There is a problem as to how these photosynthetic organisms survive the darkness of the long winter. The idea that they may turn to a heterotrophic way of life has little evidence to support it and it seems most likely that low rates of respiration at low temperatures enable reserves to be eked out (Rodhe 1955). Having low compensation points, perhaps less than

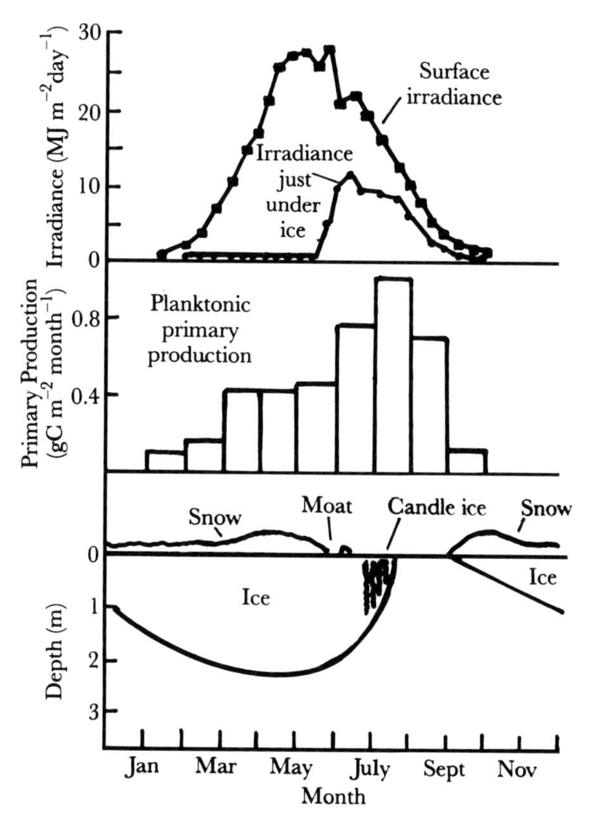

Fig. 5.1 Light conditions, planktonic primary production, and cover of snow and ice during a year in Char Lake. (Redrawn after Rigler 1978, courtesy of E. Schweizerbart'sche Verlagsbuchhandlung, Stuttgart.)

1 μmol quanta $m^{-2}s^{-1}$, they are able to carry out some photosynthesis in such dim light as may be available (see also p. 141).

Animal communities in Char Lake are simple. One copepod, *Limnocalanus macrurus*, dominates the plankton. The benthic fauna is more diverse, with about 45 species which include 24 kinds of nematodes and 7 of chironomid midges together with rotifers, tardigrades, enchytraeid worms, and other small, detritus-feeding invertebrates. For about eight months of the year 20% of the deposits which these animals inhabit is frozen, but they survive. During the summer, development of the fauna is most active in the rocky zone but this is difficult to sample quantitatively. None of these organisms is more metabolically active at low temperatures than are representatives of the same species taken from warmer sites. Adaptation to the lake environment of the high Arctic evidently consists in ability to survive rather than being particularly active at low temperatures. Mortality rates are low. Char Lake contains one species of fish, the Arctic Char *Salvelinus alpinus*, as major predator. The biomass of this fish, at 1.56 g C m^{-2}, is almost three times as large as that of all the other animals put together. However, its growth rate is extremely slow, around 0.03 $g\,C\,m^{-2}\,yr^{-1}$, and the production of invertebrate prey seems sufficient to maintain this. The Arctic char is normally an *anadromous* species (i.e. usually living in the sea and migrating into rivers and lakes to breed), but Char Lake does not have an outlet large enough to permit migration.

An Arctic lake such as this has advantages if one wishes to obtain a general quantitative picture of carbon or energy flow. The number of species is small so that food webs are simple and for a greater part of the year the system is sealed off from outside influences by ice and permafrost. For three months the lake is in darkness, a circumstance which allows it to be established that for a variety of Arctic lakes the overall respiration rate is constant from October to May. On this basis a mass balance for Char Lake can be constructed, thus:

Gains ($g\,C\,m^{-2}\,yr^{-1}$)		Losses ($g\,C\,m^{-2}\,yr^{-1}$)	
Primary production	21.1	Respiration	13.6–16.1
Allochthonous input	2.5	Output	2.1
		Storage in sediments	5.4–7.9
Totals	23.6		21.1–26.1

The amount stored in sediment is here estimated by difference. Different methods of direct measurement give values between 2 and 5.8 $g\,C\,m^{-2}\,yr^{-1}$. Even in such a relatively simple system it is difficult to obtain accurate estimates of these crucial variables. It is clear, however, that this type of lake retains primary products efficiently and respires the greater part, little being lost to the outlet or storage.

The smaller lakes on Disko Island and the adjacent mainland of western Greenland at around 69°30′N and on the east coast of Greenland between 72 and 73°N show generally similar features to those just described. The deeper lakes are oligotrophic, with sparse phytoplankton in which desmids predominate, although the flora varies from lake to lake. Shallow lakes in limestone areas, in contrast, are *eutrophic* (i.e. rich in nutrients), with pond weed, *Potomogeton filiformis*, growing on the margins and a rich fauna of chironomids. Water temperatures are higher and the length of the ice-free period longer in shallow lakes (Hansen 1967).

5.2.2 Permafrost lakes

Permafrost is most effective in lake formation in extensive deposits of unconsolidated, fine-grained, material, where it produces *thermokarst* or *thaw lakes*, unique to the Arctic. They occur in profusion on the north coasts of Siberia, Alaska, and Canada in flat lowland areas where permafrost prevents underground drainage. Soil water is not uniformly distributed (p. 69). Discrete lenses and wedges of ice form and grow by accretion. If such masses melt then a pool or lake is formed. Often, pools are connected by channels in a beaded pattern. In Alaska thaw lakes tend to be elongated perpendicular to the prevailing wind, being from several hundred metres to several kilometres in length, and arranged in a parallel fashion. They are rarely deeper than 3 metres. The mechanism producing this peculiar configuration may perhaps be subsurface currents, set up by a constant wind, eroding the shore into elliptical shape (Livingstone 1963; Sage 1986). There is a cycle as thaw lakes advance across the terrain, merge, drain, and reform in the course of thousands of years. After capture and drainage by small streams, empty basins are subject to frost action and low centred polygons with ponds in the middle are established. These may coalesce and the cycle starts again.

Some lakes in permafrost areas are larger than thaw lakes. Ozero Taymyr at about 74°N 102°30′E in Siberia, area 4650 km^2 and maximum depth 26 m, is one of the largest. It is a relict lake in a recently uplifted land surface, including a basin that was once covered by sea.

Aquatic flowering plants and mosses grow in the shallow pond margins but there is little phytoplankton. Detritus supports large populations of chironomids and other aquatic insect larvae, which provide a rich source of food for diving ducks and other insectivorous vertebrates.

Where thaw lakes are near the sea they may become saline. Imikpuk, near Point Barrow, occasionally receives seawater during the summer, then, during the winter, when the water freezes, the salt becomes concentrated almost fourfold (Livingstone 1963). This might have considerable effects on the flora and fauna.

More striking effects of salinity are seen in Garrow Lake on Little Cornwallis Island (75°23′N 96°50′W), with an area of 418 hectares, a mean depth of 24.5 m and a maximum depth of 50 m (Dickman and Ouellet 1987). It is covered for 11 months in the year with ice up to 2.4 m thick and beneath this the water was once permanently stratified (i.e. was *meromictic*). The stratification was maintained by a density difference, the water below about 15 m having 2.5 times the salinity of seawater whereas above this the water was fresh to slightly brackish. Isotope-dating showed that the bottom water had been unmixed for nearly 2500 years. The top 10 m remained free to circulate in the ice-free period. This situation seems to have come about by an advancing permafrost wedge forcing brine from rock strata below through an unfrozen chimney in the centre of the lake basin. This had a striking effect on water temperatures. The ice being clear and not usually covered with snow, it transmitted solar energy and this heated up the water at 20 m to around 9 °C, which was maintained throughout the year, although the mean annual air temperature was −16 °C and that for July only 4.3 °C. Geothermal heating was not involved. The permanent stratification had the result that oxygen was depleted from the bottom waters by decomposition of sedimenting organic matter (Fig. 5.2).

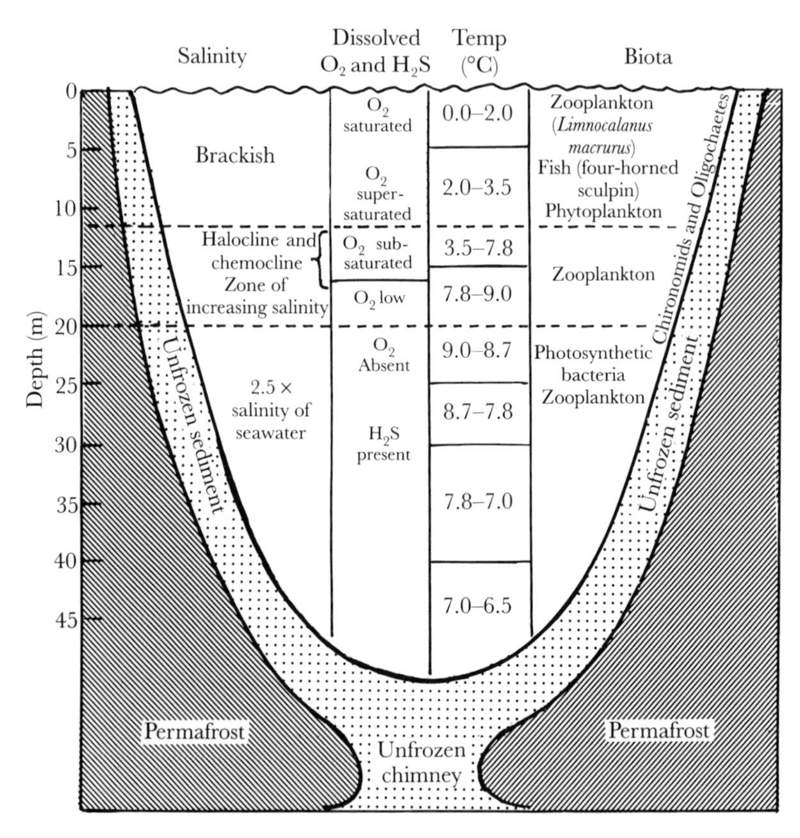

Fig. 5.2 Schematic section of Garrow Lake, Little Cornwallis Island, Northwest Territories, showing physical features and biota. (Modified from Dickman and Ouellet 1987, by courtesy of the editor, *Polar Record*.)

These physical circumstances provided a suite of different habitats. The surface water had flora and fauna rather similar to that of Char Lake. Following thaw, centric diatoms growing on the undersurface of the ice were mixed into the surface water by wind circulation but for the rest of the year diatoms were comparatively rare and microflagellates, such as *Rhodomonas minuta* and cryptomonads, dominated. The only crustacean zooplankter was *Limnocalanus macrurus* and the only fish was the four-horned sculpin *Myoxocephalus quadricornis*. The sediments in this depth zone were dominated by chironomids. A dense layer of *Chromatium*-type photosynthetic bacteria, with between 0.1 and 5×10^5 cells ml^{-1}, occurred around 20 m at the depth where oxygen disappeared from the water (Fig. 5.2). Purple sulphur bacteria like *Chromatium* require anoxic conditions and the presence of hydrogen sulphide to act as a hydrogen donor to carry out photosynthesis. The hydrogen sulphide was generated in the anoxic sediments by bacterial reduction of sulphate in decomposition processes – sulphate being abundantly present in brine derived from seawater. The photosynthetic bacteria, besides providing a significant source of food for the *Limnocalanus*, were the main agents absorbing the radiant energy which warmed the water.

This account has had to be written in the past tense. In November 1981 a mining company began discharging wastes into the lake, amounting to 100 metric tons (tonnes) of zinc and lead mine tailings per hour. Whereas little immediate effect on phytoplankton was observed, the abundance of photosynthetic bacteria declined within two years to negligible levels. Since these bacteria played key roles both in determining the physical conditions within the lake and as primary producers it must be concluded that this remarkable ecosystem has been irreversibly damaged (Dickman and Ouellet 1987).

5.2.3 Sub-Arctic lakes

Torneträsk in Swedish Lappland, 342 m above sea level at 68°20′N 19°20′E, is a lake of 317 km^2 area and 169 m maximum depth, formed by glacial excavation and partly held by morainic dams. It is extremely oligotrophic with primary production during the summer around 20 mg C m^{-2} day^{-1} but it has a longer ice-free period, about 4 months, than high Arctic lakes. Its flora and fauna are more diverse than these other lakes and it supports a fishery for Arctic char, which in this lake, unlike in Char Lake, is anadromous.

Two contrasting lakes in Iceland are Thingvallavatn and Mývaten, respectively oligotrophic and eutrophic (Jónasson 1979, 1992). Both are situated in volcanic areas with catchments characterized by basaltic lava flows and both have a geothermal input. Volcanic activity continues in the Mývatn area whereas the Thingvallavatn region is quiescent. Both lakes are fed mainly by springs and their fluxes of nitrogen, phosphorus and silicon are 1.4, 1.5, and 340, and 1.4, 0.8, and 240 g m^{-2} yr^{-1}, respectively. The slightly higher level of nutrients in Mývatn is scarcely sufficient to account for the difference in productivity. The calcium concentration in Mývatn is between 4.9 and 25.4 mg l^{-1} whereas it is 3.2–12.8 mg l^{-1} in Thingvallavatn, which would contribute something to greater productivity in the former. The crucial factor, however, is depth. Mývatn has north and south basins which have mean depths of 1.05 and 2.3 m, respectively. Thingvallavatn has a mean depth of 34 m. Shallower water results in higher summer temperatures, the annual amplitude being 18 °C, and frequent resuspension of sediment with consequently more rapid recycling of nutrients. During winter the relatively small volume of water under the ice in Mývatn becomes depleted of oxygen and anaerobic processes release nutrients which become available when the water is circulated at ice melt. In the deeper lake the maximum temperature is 10 to 11 °C, sediment does not become resuspended to any extent, and the water never becomes anoxic.

Even when investigations are conducted by the same team, it is difficult to make comparisons between species diversities in two lakes. Both Mývatn and Thingvallavatn have much greater diversity than the lakes considered previously, with about 48 species of phytoplankton, about 150 of phytobenthos, and 3 species of crustacean zooplankton. An important difference in the phytoplankton is that Mývatn has four species of cyanobacteria, including *Anabaena flos-aquae*, a nitrogen-fixer which becomes dominant in summer and contributes appreciably to the fertility of the lake. Thingvallavatn has only two species of cyanobacteria in the plankton, neither of them abundant, but although there are vigorous growths of *Nostoc* sp., another nitrogen-fixer, at 1 m depth in the littoral zone, this can be of little importance for the lake as a whole. Macrophytes, diatoms, and the green alga *Cladophora*, which favours shallow eutrophic waters, are the dominant benthic forms in Mývatn.

Mývaten is the richer in zoobenthos, with some 229 species as against 60, and has three fish species as against two in Thingvallavatn. The production of chironomids is impressive – in still weather, swarms of adults rise from the lake margins like plumes of smoke and thick scums of exuviae accumulate on lee shores. Two of the 20 species, *Tanytarsus gracilentus* and *Chironomus islandicus*, are dominant, comprising 90% of its mean annual production. Chironomid larvae are a principal food for fish and are also important for some species of diving ducks. There are large fluctuations in chironomid numbers from year to year and the nutritional state of Arctic char seems to be related to these. Mývatn has exceptional populations of water fowl with 17 nesting species; and ducks play a significant part in its energy flow. The lake has now been designated as a conservation area by the Icelandic government.

Thingvallavatn has primary production of about 95 $g C m^{-2} yr^{-1}$ in phytoplankton and 55–118 $g C m^{-2} yr^{-1}$ in the phytobenthos whereas the corresponding values for Mývatn S. basin are 118 and 272. Thingvallavatn is about ten times as productive as Char Lake whilst Mývatn is only somewhat less productive than temperate eutrophic lakes (e.g. Loch Leven is Scotland at 56°N). In both Thingvallavatn and Mývatn the energy flows via plankton are much less than that via benthos but the efficiencies of the two pathways are about the same. Thingvallavatn is more efficient in producing fish than is Mývatn, in which ducks are dominant predators. In neither lake has energy storage in sediments been quantified (Table 5.1; Jónasson 1979, 1992).

5.3 Antarctic lakes

These fall broadly into two categories. There are numerous small lakes and ponds on the peri-Antarctic islands and around the coast of the continent whereas in the second category are others, some of moderate size and permanently ice-covered, in the dry valley areas. Except in the mild-polar zone, almost all are in basins produced by glacial action and some are ice-dammed (Heywood in Laws 1984; Vincent 1987).

5.3.1 The lakes of Signy Island

Signy Island has a number of lakes, the largest of which is 4.2 hectares in area and the deepest with a maximum depth of 15 m, lying in valleys and depressions in the narrow coastal lowland (Fig. 5.3). Typically, the lake basin is a steep-sided trough surrounded by a shelf, usually at about 1 m depth, formed by moraine damming. Snow, which varies greatly from year to year, has an important effect on the lake environment since it determines the depth to which the water freezes (1–2 m), the duration of the ice cover (8–12 months), and the irradiance immediately under the ice (0.1–20% of incident visible light). Maximum temperatures in the summer range between 1 and 6 °C when the water column is isothermal due to stirring by the strong winds which prevail. An inverse temperature gradient develops under the ice in winter, with the bottom layers reaching about 1 °C. Because a large proportion of the water turns to ice, dissolved solids are concentrated up to twice their summer levels.

Moss Lake, occupying a cirque basin of maximum depth 10.4 m in a small catchment area of rock, scree, and small but permanent patches of snow and ice, is one of the most oligotrophic of these lakes, precipitation being its major source of mineral nutrients. The water is clear and phytoplankton sparse, chlorophyll *a* concentrations varying from 0.5 to 8.0 mg m^{-3} – somewhat more than in Char Lake. Primary productivity is correspondingly low, 12.6 g C m^{-2} yr^{-1} being an estimate for one particular year. Zooplankton is also sparse and one of the more abundant cladocerans, *Alona rectangula*,

Fig. 5.3 The east coast of Signy Island. The nearer of the two lakes in the centre of the picture is Heywood Lake. Its proximity to the sea should be noted. (Photo G. E. Fogg.)

although an active swimmer is mainly benthic. Benthic vegetation is limited in depths of less than 5 m because of ice scour but the shelf has the olive-green cyanobacterial mat, with a diatom epiflora, which is common to nearly all the Signy lakes. By absorption of radiant energy these mats may achieve temperatures of 2 °C or so above the ambient water in summer. Visible radiation in the mat is attenuated to about 1% of surface intensity at 1 mm, the thickness being about 3 mm. Consequently, photosynthesis is confined to the top millimetre and below this anaerobic conditions tend to prevail, especially in winter. Heterotrophic activity in the mat is high. In deeper water there are patches of the cyanobacterial association, *Tolypothrix-Plectonema*, which is probably nitrogen-fixing. The bottom of the lake, which is covered by fine, unconsolidated, glacial sediment, is dominated by the aquatic mosses, *Calliergon sarmentosum* and *Drepanocladus* sp. Their stems may be as long as 40 cm and in shallow water they occur at a density of about 24 000 m^{-2}. This probably represents 20–50 years growth, the luxuriant appearance reflecting more a slow rate of decay than vigorous growth. However, comparison of this with the even slower growth of adjacent terrestrial mosses underlines the point that desiccation is a major limiting factor for land plants in the Antarctic. These mosses support a complex community of epiphytic algae and invertebrates, which differ according to the moss species. The epiphytes include cyanobacteria, *Oedogonium*, and diatoms. The fauna, mainly of opportunistic grazers, has sessile and swimming rotifers and cladocerans as its most numerous members, together with ostracods, tardigrades, nematodes, and gastrotrichs. This epiphytic mass reduces the light received by the moss plants. There are no fish in Moss Lake. The water remains oxygenated throughout its depth during the period of ice cover. Anaerobic bacteria are virtually absent from its sediments and there is little seasonality either in total bacterial numbers or microbial activity as measured by uptake of oxygen or dark fixation of radiocarbon-labelled carbon dioxide, glucose, or acetate. It seems that organic production in Moss Lake is largely consumed by decomposers and there is little storage of organic material in the sediment (Heywood in Laws 1984).

Heywood Lake, although adjacent and physically similar to Moss Lake, is different principally because its catchment, largely moss-covered, is accessible to seals which contribute considerable amounts of organic matter to the water. The activities of these animals have transformed the lake from an oligotrophic to a mesotrophic condition within the last 30 years. The water is turbid in summer, horizontal visibility being reduced to about 20 cm, but it becomes clear in winter under ice. The relatively abundant phytoplankton, which includes small species of *Ochromonas*, *Cryptomonas*, *Chlamydomonas*, and *Ankistrodesmus*, reaches a peak in the spring. A cryptophyte, *Rhodomonas minuta*, may become dominant later but diatoms are rare. It is striking that, as in other Antarctic lakes, desmids are virtually absent whereas they are frequent in Arctic freshwaters. Another notable absence is that of planktonic cyanobacteria. A chlorophyll *a* concentration of 170 mg m^{-2} has been recorded for the spring population and primary production sometimes reaches 3 g C m^{-2} day^{-1} on the rare sunny days. In 1970–2 the assimilation number (the ratio of carbon fixation per hour at light saturation to chlorophyll concentration) was 10.5, a high value, about ten times that in an adjacent sea area, indicating phytoplankton adapted to high light intensities. In a shallow, well-mixed, lake this is to be expected. Chlorophyll concentrations fall rapidly when the lake freezes and few algal cells are then to be found in the water. The annual primary productivity of the phytoplankton has been estimated as 139 g C m^{-2} (Table 5.1). The main rocky shelf of the lake is covered with a *Phormidium* mat similar to that of Moss Lake but the deeper parts are nearly bare with just a film of *Phormidium* and rare tufts of moss on the slopes. A crustacean, *Branchinecta gaini*, dominates the zooplankton in summer, early larval forms feeding off the phytoplankton and perhaps accounting for 40% of the spring bloom in two weeks. Late larval stages and adults feed on epilithic algae and detritus. The invertebrate benthos has fewer species than are found in the oligotrophic lakes. No turbellarians or gastrotrichs have been found and the number of species of rotifers and crustaceans is reduced. Numbers of individuals, on the other hand, are greater. Increased productivity but decreased diversity is a familiar effect of organic enrichment of a lake and is often taken as an indication of pollution.

The water of Heywood Lake becomes anoxic under ice cover, which lasts between 8 and 10 months. The sediment is anaerobic and about 10% of its dry weight is organic. It contains significant populations of facultatively anaerobic heterotrophic and sulphate-reducing bacteria. There is seasonal periodicity in bacterial numbers with a peak in late autumn and a winter minimum in August. Aerobic bacteria contribute largely to the total and mask the counts of viable anaerobes, which follow an inverse pattern with maxima at the height of winter. Oxygen uptake by the sediment parallels total bacterial numbers, becoming almost undetectable in winter. Uptake of radiocarbon-labelled carbon dioxide, a measure of chemosynthetic activity, on the other hand, reaches its peak in winter (Ellis-Evans 1982).

The total annual carbon gain in Moss Lake is less than an eighth of that in Heywood Lake. However, whereas the total gains and losses approximately balance in Moss Lake, there is a great discrepancy in Heywood Lake, with more than half of the input of carbon unaccounted for. There is uncertainty about the input of allochthonous organic material and loss in the outflow, neither of which have been quantified, but probably the major part of the missing carbon is stored in the sediment of Heywood Lake.

5.3.2 The lakes and ponds of the McMurdo Sound area

Cape Evans, a low ice-free area of black basaltic lava on Ross Island at 77°38′S 166°25′E, has a number of small lakes with diverse biological characters. Skua Lake is a favourite haunt of the local birds and Algal Lake was so called because of a conspicuous mat of cyanobacterial remains on its leeward side. Coastal pools receive sea spray and become saline to varying degrees. Pony Lake, near the Cape Royds Adélie penguin rookery (77°33′S 166°08′E), has concentrations of ammonium ion which are sometimes more than 5000 $\mu g l^{-1}$ and soluble reactive phosphorus greater than 2000 $\mu g l^{-1}$. Its phytoplankton biomass may rise to as high as 347 mg chlorophyll *a* m^{-3}. All these lakes freeze almost solid during the winter but a residuum of saline water at the bottom becomes concentrated more than 15-fold so that it remains liquid even at −13 °C. Thus, the physical and chemical characteristics of these water bodies are highly unstable and the duration of conditions suitable for plant and animal growth is brief. An endemic rotifer is well adapted to such conditions, being reported as able to survive in its dormant state salinities up to 250‰ and temperatures down to −78 °C.

In this region, which has more hours of sunshine than Signy Island, there is strong inhibition of phytoplankton photosynthesis by high light intensities. Algal and Skua Lakes both show periodicity in carbon fixation out of phase with the diurnal variation in radiation, maximum rates of photosynthesis being found at midnight (Fig. 5.4). This effect is more pronounced near the water surface (5 cm) than at a depth of 50 cm, and in the more productive Skua Lake, in which phytoplankton is self-shading, than in the clearer Algal Lake. When samples are shaded by neutral filters maximum photosynthesis occurs at 20% of the incident solar radiation at noon. Rise in temperature increases the rate of carbon dioxide fixation in photoinhibited samples with a Q_{10} of about 7 whereas similar but non-inhibited samples have an average Q_{10} of about 2 (p. 33). Photoinhibited algae recover if kept in dim light for a few hours. Since experiments have been conducted in glass bottles, opaque to UV radiation, these effects must be produced by visible, photosynthetically active, radiation (and these experiments were done before the advent of the 'ozone hole'). Although the phytoplankton of the two lakes, differing markedly in transparency, standing crop, and rates of production, show large differences in extent of inhibition, their responses in terms of overall ecological efficiency (energy fixation by phytoplankton as a fraction of incident radiant energy) are similar. Higher biomass-related production in Skua Lake is offset by its greater sensitivity to high light intensity and by the greater depth of water column available for photosynthesis in the clearer Algal Lake (Goldman *et al.* 1963).

5.3.3 The dry valley lakes

In the dry valley regions there are lakes frozen solid to the bottom, others permanently ice-covered and meromictic, and some which are so saline that they cannot freeze (Vincent 1987; Green and Friedmann 1993).

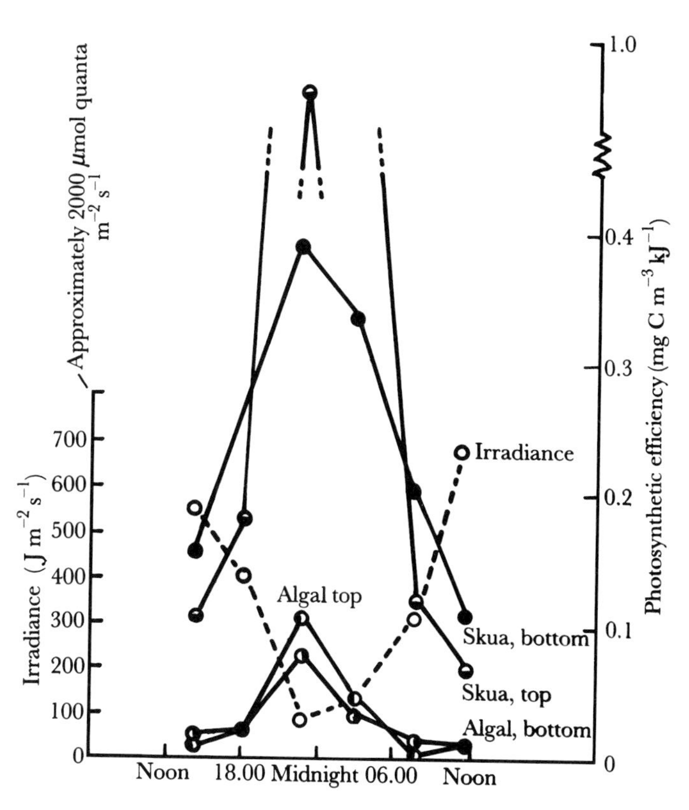

Fig. 5.4 Variation over 24 hours in efficiency of photosynthesis, as determined by the radiocarbon method, in two small lakes, Skua (turbid with phytoplankton) and Algal (clear water) at 77°38′S 166°24′E. Marked reduction was shown at noon and maximum efficiency, both at the surface and at the bottom, occurred at midnight. Irradiance is given in joules but the approximate equivalent in photon flux density is indicated. It should be noted that reduced efficiency is caused by a combination of photoihibition and irradiance above the saturating level. (Modified from Goldman *et al.* 1963, courtesy of the editor, *Limnology and Oceanography*.)

Lake Vida (77°24′S 162°0′E), 5 km long, is presumed to be frozen through to its base. Unsurprisingly, there seems to be nothing known about its biology. With a mean air temperature in this region of −20 °C one may wonder how any lakes are able to retain any liquid water at all. Partly, this is because lakes lying over permafrost act as heat sinks, collecting 'hot' run-off water which is then protected from wind disturbance by a 3 metre-thick permanent ice cover. For a given air temperature regime this thickness is fairly uniform from lake to lake because each winter a new layer of ice is added to the bottom of the cover and the latent heat released is about equivalent to the heat loss from the ablating ice surface. However, in Lake Vanda (77°32′S 161°33′E), 5.2 km^2 in area and maximum depth 66.1 m, another factor, which we have already seen operating in the Arctic (p. 113), comes into play. There is only one inflow, the Onyx River, and the heat input from this is too limited in duration to affect the temperature of the lake significantly. There is no outflow from the lake. The main source of thermal energy is solar radiation pentrating the exceptionally transparent ice and clear water during the 24-hour days of summer. The vertical turnover of the ice constrains its crystals to grow in the vertical direction, forming optical pipes transmitting radiation into the water. When the ice thaws around the lake margin these crystals are released as 'candles' (Fig. 5.5). The water is strongly stratified with a saline layer at the bottom so that heat is not dissipated by convection and the bottom water reaches about 25 °C–46 °C above the mean temperature in the valley. It was once thought that this must be due to geothermal heating but a hole drilled through the 12 m of sediment showed a heat flux from the water into the sediments, rather than the other way round. Sandwiched in between the stable density gradients in the water column are thermohaline convection cells within which there is rapid horizontal transport. A numerical model with solar radiation as the forcing function accurately describes this thermal structure. The origins of the meromictic condition are not clear. Possibly there have been large fluctuations in lake area and depth during the last few thousand years and, after a period of salt concentration by ablation of ice cover during a colder period, the River Onyx resumed its flow and overlaid the brine with freshwater.

Logistics restrict access to Lake Vanda except in a brief November–January field season but observations starting in mid-September have recently been made in adjacent Lake Bonney. At this time the irradiance immediately below the ice is only 0.03 mol quanta $m^{-2} day^{-1}$ as compared with 1.9 in

Fig. 5.5 Candle ice breaking up, Sombre Lake, Signy Island, January. (Photo G. E. Fogg.)

mid-November. Nevertheless, photosynthesis begins in early September and the maximum for phytoplankton biomass and production moves progressively down the water column, following the seasonal increase in irradiance (Lizotte *et al.* 1996). Whereas the dry valley lakes usually have relatively high concentrations of inorganic nitrogen, they are poor in phosphate and Lake Vanda is one of the most phosphorus-deficient and oligotrophic freshwaters known. The phosphorus deficiency may result from its removal by abiotic formation of the mineral hydroxyapatite in the bottom waters. Through most of the water column the chlorophyll *a* concentration is extremely low, around 0.1 mg m^{-3}, but it reaches a maximum of 0.6 mg m^{-3} at 58 m, the boundary between oxygenated and anoxic waters (Fig. 5.6). Photosynthesis is light-limited rather than temperature-limited and its efficiency is low, about 0.1 mg C mg chlorophyll a^{-1} h^{-1} (W. F. Vincent and C. L. Vincent 1982). The planktonic primary producers are mainly flagellates, which, being able to swim, can accumulate at the depth most suitable for their growth, or small non-motile forms which sink very slowly. This results in stratification of the communities, each adapted to the light, temperature, and chemical conditions in their respective layers. Three floristically distinct communities have been found. Microflagellates, including *Ochromonas miniscula* and *Polytomella* sp., occur just under the ice. They benefit from the rather higher nutrient levels resulting from solute exclusion

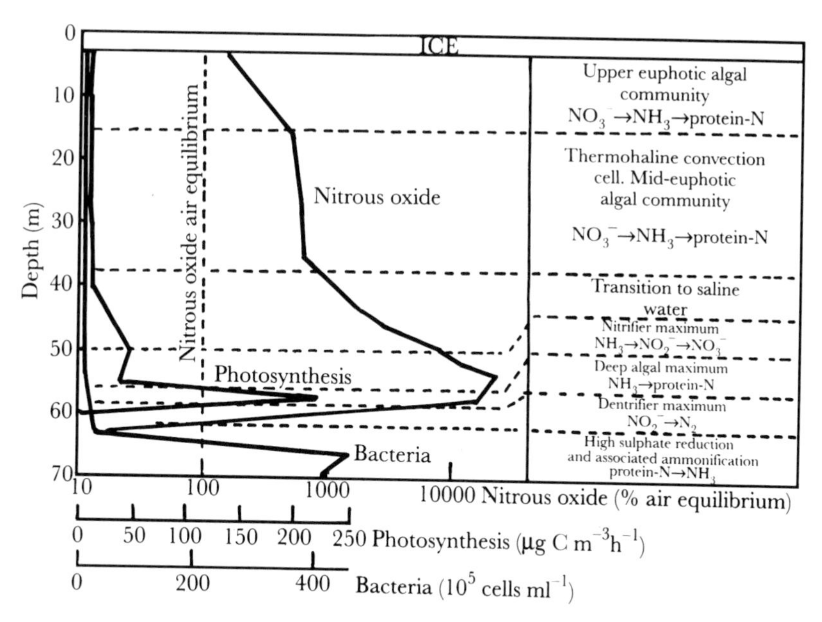

Fig. 5.6 Schematic section of Lake Vanda, Wright Valley, southern Victoria Land, showing stratification of different processes in nitrogen cycling. The various microbial components are located in specific depth zones. The transition from oxygenated to anoxic water at around 58 m is critical. (Data of Vincent 1987, courtesy of the author and the manager, SIR Publishing, Wellington, New Zealand.)

as fresh ice is formed. Nitrate, for example, is between 50 and 100% higher at 5 m than it is at 30 m. This community seems to be adapted to the low irradiances of spring, when most of its growth evidently takes place, rather than to relatively higher summer values. A second community, with *Phormidium* spp. as well as flagellates, exists in convection cells between 15 and 38 m. The third community, responsible for a deep chlorophyll maximum at 58 m, is dominated by two cyanobacteria, *Phormidium* sp. and a *Synechocystis*-like form, adapted to warmth and dim light. It must suffer severely from phosphorus deficiency since the nitrogen : phosphorus ratio is greater than 5000 : 1 (in seawater it is usually 16 : 1). In the oxygenated water the distribution of heterotrophic bacteria follows approximately that of the algae but their numbers rise steeply in the anoxic bottom water. Different biochemical types of bacteria show a well-defined layering, reflected in the vertical distribution of nitrous oxide, an intermediate in nitrification (oxidation of ammonia to nitrate). In the upper community, where nitrate is reduced to ammonia and assimilated into protein, its concentration is in equilibrium with that in air (Fig. 5.6). In the deep chlorophyll maximum, where ammonia diffusing up from the anoxic layer is nitrified, it increases to over 20 000% of the air equilibrium concentration. Below this, sulphate reduction is the dominant process, nitrate is removed by denitrification, ammonia is produced by degradation of protein, and the concentration of nitrous oxide falls to less than the air equilibrium value. The zooplankton in Lake Vanda is extremely sparse, comprising one or two ciliates, such as *Strombidium* spp.; crustacea, aquatic insects, and fish are absent. The total plankton biomass has been put at 45 $\mathrm{mg\,C\,m^{-3}}$ and its production, at between 0.34 and 2.5 $\mathrm{g\,C\,m^{-2}\,yr^{-1}}$, is among the lowest recorded.

The littoral zone of Lake Vanda, about half of its area, is covered by algal mats. As elsewhere, these are composed principally of cyanobacteria with some diatoms and a fauna of ciliates and rotifers. These mats assume different forms which Captain Scott and his companions glimpsed through the ice of Lake Bonney in 1911 and described just as 'water plants'. Besides the usual prostrate type there are 'lift-off' and 'pinnacle' mats. Pinnacle mats have conical protrusions, 2–5 cm in height, which in cross-section show layers of light brown to tan zones alternating with greenish-purple layers and commonly with sand or calcite crystals included (Simmons *et al.* in Green and Friedmann 1993). Pinnacle mat is often associated with a moss resembling the usually terrestrial species, *Bryum algens*. Although this is the most widespread moss south of 56°S it is absent from adjacent dry valley lakes and also from the summer ice-free moat around the edge of Lake Vanda itself. This puzzling distribution has not been explained. The primary productivity of algal mats in Lake Vanda does not seem to be known but in nearby lakes estimates of biomasses of 8–480 $\mathrm{mg\,C\,m^{-2}}$ yielding up to 113 $\mathrm{g\,C\,m^{-2}\,yr^{-1}}$ have been made. As in Char Lake in the Arctic, benthic production is distinctly higher than that of the phytoplankton. Bubbles formed by photosynthesis may cause pieces of mat to float up, detach, and become embedded in the ice. As new ice accumulates below and ice above is lost by ablation, these pieces work their way up and, after 5–10 years, reach the surface and are blown away. This may be an important means of dispersal of organisms and also a significant drain on the nutrient resources of the lake.

Lake Fryxell in the Taylor Valley is generally similar to Lake Vanda but much more productive with a plankton biomass of sometimes more than 25 mg chlorophyll *a* m^{-3}. This is probably related to a better supply of phosphate, the nitrogen : phosphorus ratio being around 4 : 1. Its ice is less transparent than that of Vanda and for this reason the lake does not heat up so much, reaching a maximum temperature of 2.5 °C at 9 m, midway down the water column. Salinity increases from freshwater beneath the ice to 10‰ at the bottom. Calcite deposition, induced by photosynthetic activity at the phytoplankton maximum, occurs. Again there is a structured suite of communities of microorganisms. An algal peak deep down where the water is becoming anoxic is composed mainly of *Pyramimonas* sp. and *Chroomonas lacustris*. Relative to those of other heterotrophic protozoa and bacteria, numbers of ciliates are high. Compared with other lakes in the region, both the plankton and the benthic communities have low species diversity. Calcite layers in the cyanobacterial mats give them a resemblance to *stromatolites*, algal formations which occur in tidal zones in tropical seas at the present day and as fossils back to the Precambrian era.

The nearly self-contained nature of the ecosystem, the unusual patterns of mineral cycling arising from low diversity and biomass of consumer animals, as well as the different biochemical regimes maintained in discrete layers in the water column, make Lakes Vanda and Fryxell, and others like them in southern Victoria Land, biologically unique. A different type in the same region is represented by Don Juan Pond, about 300 by 100 m, and only about 10 cm deep. Its interest is that it is a nearly saturated solution of calcium chloride, with the highest known salinity in Antarctica. It remains unfrozen in temperatures which go down to −50 °C, the calculated freezing point of its waters being −54 °C. The presence of life in it is debatable. Several bacterial species and a yeast have been reported but recent investigators have failed to confirm the isolations and it seems likely that the organisms were washed in by a temporary and relatively dilute inflow (Vincent 1987).

5.3.4 Vestfold Hills lakes

The Vestfold Hills, extending from latitudes 68°25′S–68°40′S and longitudes 77°50′E–78°35′E, present an ice-free area of about 500 km^2 with several hundred lakes and pools lying in depressions produced by glacial action. Remains of many types of marine organisms in the mineral soil show that the area was under the sea before the glacial retreat of about 6000 BP. Hypersaline lakes were created at that time from trapped seawater. Many are fed by small melt water streams but none have outlets. Some still connect with the sea on some tides. Dry katabatic winds concentrate the waters, resulting in some lakes having salinities four times that of seawater or more. Such lakes do not freeze although the annual mean temperature is −10 °C and the lowest monthly mean temperature −18 °C.

Ace Lake has characteristics generally similar to those of the southern Victoria Land lakes, except that the ice, if present, usually disappears during the months of January to March (Vincent 1987). However, it has some features which have not been studied in lakes elsewhere. The phytoplankton flora is remarkably simple, having only four recorded species, all flagellates. They are *Pyramimonas gelidicola*, *Cryptomonas* sp., and, less abundant, an unidentified microflagellate, and a dinoflagellate. These are distributed in a vertical pattern, as in Lake Vanda, but here it has been shown to be a dynamic stratification. There is no seasonal succession of species but the four flagellates shift their positions with the changing light regime. From an irregular distribution in winter, initial spring growth concentrates in the 2–5 m layer beneath the ice. With improving irradiance the flagellate maxima begin to move downwards. The *Cryptomonas* population sinks to 5 m by late October and stays there whereas the *Pyramimonas* peak moves to 10 m, where oxygen begins to be depleted and nutrients are in good supply but irradiance low, by January. The unidentified flagellate, however, remains in the surface 2–5 m. *Pyramimonas* and the unidentified flagellate are most abundant in the open water in February but conditions then evidently do not suit *Cryptomonas*, which has its maximum in November.

Large numbers of photosynthetic bacteria, *Chromatium* and *Rhodospirillum* spp., about $3 \times 10^{7}\,ml^{-1}$, are active at the top of the anoxic layer. Their metabolism seems closely linked to that of the anaerobic sulphate-reducer, *Desulfovibrio*, the phototrophs providing organic carbon to the sulphate-reducer which in turn provides the sulphide necessary for photosynthesis. Although the ionic composition of Lake Ace water should reflect that of seawater, the content of sulphate is only 24% of the expected value. The deficit probably arises via reduction of sulphate to sulphide. This agrees with the present proportion of the heavy isotope of sulphur, ^{34}S, being well above that of seawater – the lighter isotope, ^{32}S, would be preferentially removed by reduction of sulphate to sulphide. Sulphide would diffuse upwards and be utilized by phytoplankton after reoxidation to sulphate. Sulphate is not volatile but many algae are known to convert it into the volatile compound dimethyl sulphide (p. 188) which escapes from the water into the atmosphere carrying a higher proportion of the ^{32}S with it (Vincent 1987).

Ace Lake has benthic algal mats similar to those already described. Animal life includes the calanoid copepod *Paralabadocera antarctica*, which feeds on phytoplankton. This species also occurs in inshore waters of the Southern

Ocean and there may well be other elements of an original fjord fauna in the Vestfold Hills lakes.

5.3.5 Epishelf lakes

In coastal regions there exists a rare type of lake, contained by ice but showing changes in level over semi-diurnal periods, which evidently has some connection with tidal waters. The first to be found was in the Schirmachervatna (70°S 11°E) but the best known is Ablation Lake at 70°49′S 68°25′W on the west coast of George VI Sound, separated by 100 km of shelf ice 100–500 m thick from the open sea (Heywood 1977). The lake, 6.5 km^2 in area and more than 70 m deep, is held back by the shelf and is permanently covered by ice varying from 2.5 to 3 m in thickness in summer and 4 to 4.5 m in winter. This moves up and down with an irregular rhythm through a maximum recorded range of 1.65 m. Only 15–20% of incident radiation penetrates the ice when it is snow-free but it is usually snow-covered. The area has a polar night effectively lasting two months because of adjacent mountains and so the annual input of solar energy to the lake must be low. The water is fresh down to 55 m and the seawater below it, at −2 °C, acts as an effective heat sink so that isothermal conditions at about +0.1 °C usually prevail in the upper layers. Below the halocline salinity increases to reach about the usual value for seawater at 66.5 m. Variations in the salinity profile with tidal movement indicate that the saline layer is in direct contact with the waters of the Sound.

Ablation Lake contains both freshwater and marine species. The only benthic vegetation is a thin film of cyanobacteria on a few rocks sticking out from the floor of firm silt in the shallows. Planktonic chlorophyll *a* is 0.5–0.65 mg m^{-3} (i.e. somewhere between concentrations in Lakes Moss and Vanda). A maximum photosynthetic rate of 60 mg C m^{-2} day^{-1} is recorded. The fauna of the freshwater stratum includes *Pseudoboeckella poppei*, a non-marine copepod found in other lakes in the region, but rotifers, tardigrades, and nematodes have not been recorded. An unidentified copepod occurs at depths below 40 m in water of 0.6 to 32‰ salinity. In the seawater layer there is the common Antarctic marine fish, *Trematomus bernacchi*. Fish sampled have been in good condition and have recently fed on a variety of marine organisms. This species is not adapted to freshwater and individuals brought to the surface of the lake show distress. Since it is of sedentary habit its presence indicates that there is normally a marine community at the bottom of the lake.

5.3.6 The lakes of the sub-Antarctic islands

Small lakes and ponds occur in abundance on the islands of Macquarie, Kerguelen, Marion, and South Georgia, but few have been investigated in detail. On Macquarie Island ponds fill depressions in waterlogged peat and lakes of medium size – less than 1 km^2 in area and between 0.6 and 32.3 m deep – occur on tundra soils or fill rocky basins (Selkirk *et al.* 1990). The waters receive most of their ionic constituents from atmospheric precipitation, with a strong marine contribution from the incessant westerlies. The ponds on peat have a tinge of humic coloration. Humic substances absorb strongly in the UV radiation and thus afford protection to phytoplankton. Complete ice cover is rare and in summer mean water temperatures may get as high as 9 °C. The lakes are oligotrophic with relatively sparse vegetation, although this shows more species diversity than the Antarctic lakes. The aquatic angiosperm *Myriophyllum triphyllum* forms dense stands down to about 2 m. Two other aquatic flowering plants and some five submerged mosses are also found. The plankton comprises green algae and diatoms. The chlorophyll *a* content of the water varies between 1.5 and 24 mg m^{-3}. The circumpolar sub-Antarctic copepod, *Pseudoboeckella brevicaudata*, is the most abundant species in the zooplankton and several other crustaceans, together with oligochaetes, nematodes, flatworms, rotifers, tardigrades, and insect larvae, occur either in the plankton or the benthos. There are no fish. The other sub-Antarctic islands likewise have no indigenous freshwater fish but salmonids were introduced into Kerguelen rivers and lakes from 1954 onwards and brown trout (*Salmo trutta*) and brook trout (*Salvelinus fontinalis*) have established themselves successfully. An attempt to introduce rainbow trout (*Salmo gairdneri*) into a South Georgian lake failed.

5.3.7 Volcanic lakes

Some lakes and pools in the Antarctic, as in the Arctic, have been formed or are affected by volcanic activity, Gorgon Pool on Candlemas Island (57°05′S 26°41′W, Heywood in Laws 1984), for example. The only one which has attracted even casual interest from limnologists is Kroner Lake on Deception Island (63°00′S 60°34′W). This occupied a shallow circular depression in a lava plain. Fumarole activity suppied heat so that the lake never froze and a maximum of 10 °C was sometimes reached. Because of its proximity to the sea its water was brackish and the most obvious vegetation comprised brown and red seaweeds, although it does not seem to have been recorded whether these were attached or blown in from the seashore. In a bacteriological study, Gram-negative rod-shaped bacteria, mostly with optimum temperature for growth below 20 °C, were found to predominate. They evidently utilized decomposition products from the macroalgae. In view of the production of hydrogen sulphide by sulphate reduction accompanying decomposition of seaweeds and the presence of this gas in the spring waters, photosynthetic bacteria would be expected to have been abundant but do not appear to have been recorded. The lake disappeared in 1969 following a volcanic eruption and a basin connected to the sea by a narrow channel now occupies its place. This new Kroner Lake (Fig. 5.7) has been declared a Site of Special Scientific Interest but its attraction as a harbour for small boats has meant that the protection afforded by this declaration is sometimes violated.

Fig. 5.7 Kroner Lake, Deception Island, January 1993. The connection between the lake and the sea is visible on the far side just left of centre. The light zone, yellowish in colour, on the nearer shore shows the presence of sulphur bacteria. (Photo G. E. Fogg.)

5.4 Streams and rivers

As in other parts of the world, the streams and rivers of polar regions have taken second place to lakes with limnologists.

5.4.1 The Arctic

No major rivers have their sources in the Arctic but about 14%, more than 10 million km^2, of the land area of the world drains into the Arctic Ocean via five of its largest rivers – Yenisey/Angara, Ob/Irtysh, Lena, Kolyma, and Mackenzie. These discharge through deltas which usually present a maze of channels and shallow lakes. The effect of this inflow of relatively warm, fresh, water into the Arctic Ocean has already been touched on (p. 8).

The Mackenzie River, draining nearly a fifth of the total land area of Canada, has a flow which is more evenly distributed through the year than that of smaller rivers with catchment areas, mostly tundra, with run-offs having a peak at the spring melt. The Colville River in Alaska had 43% of its discharge in 1962 concentrated in a three-week period. Such spates make the substratum highly unstable. Where high-energy streams flow through non-cohesive deposits they carry heavy loads and develop braided multiple channels, constantly shifting in position. The fine sediments are often colonized by tundra vegetation, which is particularly rich and productive in such a situation. For some six months in the year Arctic rivers and streams are ice-covered and their huge deltas are frozen to depths of several metres. The cataclysmic break-up of the ice in the spring gives overwhelming erosive power (Sage 1986). In all, these rivers do not provide habitats favouring the establishment of either plankton or stable submerged communities. In any case, most of these rivers have their sources in latitudes well south of the tree line and flow through regions which are non-Arctic in character for much of their lengths. Water temperatures are higher than those of the surrounding terrain and both benthic and pelagic species would tend to be those of temperate latitudes.

The classification of Alaskan streams and rivers by Craig and McCart (1975) is applicable generally in the Arctic. The three types distinguished are as follows. (1) *Mountain streams*: fed by springs and surface run-off; waters rarely exceeding 10 °C; flow for about 5 months in the year; density of benthic invertebrates low, of the order of 100 organisms m^{-2}; Arctic char the common fish species. (2) *Spring streams*: small, spring-fed, tributaries of mountain streams providing a more stable habitat; mean temperatures 2.5 °C in winter and 7 °C in summer; banks often overgrown with vegetation and beds largely covered with moss or algae; high densities of benthic invetebrates, around 10 000 organisms m^{-2}; Arctic char the main fish. (3) *Tundra streams*: draining the peat of foothills and coastal plains; tend to be small and meandering, flowing erratically for 3.5–4.5 months in the year; usually 'beaded' with alternation of pools and riffles, the pools liable to be isolated in dry periods and grading into the static pools discussed earlier (p. 113); waters more acid, with less calcium in solution than in mountain and spring streams, stained with humic materials; little organic matter in suspension but much, up to $14\,mg\,l^{-1}$, in solution, exporting around $3\,g\,C\,m^{-2}$ annually from the drainage basin (Oswood *et al.* in Reynolds and Tenhunen 1996); summer temperatures rising to over 16 °C; densities of benthic invertebrates intermediate, around $1000\,m^{-2}$, and, conforming to the pattern for all three types of stream, densities are inversely proportional to stream discharge; grayling (*Thymallus arcticus*) use these streams for spawning and are the characteristic fish.

The ecology of the sub-Arctic River Laxá in Iceland is dominated by downstream movement of detritus and animals from the eutrophic Lake Mývatn. It supports abundant salmon and trout (Jónasson 1979).

5.4.2 The Antarctic

The Onyx River flows into Lake Vanda in the Wright Valley in southern Victoria Land. With a length of 40 km it is Antarctica's longest river. Its annual discharge varies from 15 million m^3 to zero. Cloud cover immediately stops its flow and consequently day-to-day variations in discharge are enormous (Vincent 1987).

Cyanobacterial mats are the most conspicuous vegetation in Antarctic streams. The black, mucilaginous, nitrogen-fixing species, *Nostoc commune*, forms luxuriant mats in regions of low flow. Pink, orange, or olive-green mats composed mainly of *Phormidium* spp. and other filamentous cyanobacterial species are more common. They are sometimes more than 3 mm in thickness. Under the tough mucilaginous surface layer the filaments tend to occur in sheets which can sometimes be peeled apart but the undersurface has a looser structure and holds sand grains. Additionally, there may be epilithic films or crusts of other cyanobacteria and diatoms. These mats and films contain heterotrophs, including fungal hyphae as well as bacteria. The green alga *Prasiola calophylla*, which lacks the sunscreening carotenoids found in algae growing in exposed situations, occurs in cracks and under overhangs of rocks. Protozoans, nematodes, rotifers, and tardigrades may be locally abundant but there are no insect or crustacean grazers and the total biomass of secondary producers is small.

The algal mats found in streams are organized as photosynthetic tissue like those in pools and lakes (Vincent *et al.* 1993). The surface layer is carotenoid-rich and provides protection against high levels of radiation and, by absorbing radiant energy, warms the mat. This layer has only low photosynthetic activity. Below is a stratum in which concentrations of chlorophyll and the accessory photosynthetic pigment phycocyanin, reach a maximum. Here, the algae are in an orange-red shade environment, optimal for their requirements. If the mat is shaded, the cyanobacteria, which are motile, may in some cases move up to the surface in 2 hours or so, bringing about a change in colour of the mat. The strong self-shading within the mat makes it behave as a whole as a shade-adapted photosynthetic system.

Growth of the mats is limited by water rather than by nutrients. Although nutrient levels in the surrounding water are low, both ammonium ion and dissolved reactive phosphorus reach concentrations over 100 mg N or P m^{-3} in the interstitial water of the mat. Possibly, the mat acts as a nutrient-accumulating system in which there is recycling by heterotrophic components. Supply of extra nutrients does not enhance growth of the mats.

The resistance of mats to desiccation and their rapid resumption of activity on rewetting has already been discussed (p. 42). During the winter mats remain as freeze-dried crusts which experience temperatures down to −60 °C but nevertheless retain viable cells from which the next season's growth can start. Biomass levels are high, in the range 15–50 mg C cm^{-2} for *Nostoc* and 5–15 mg C cm^{-2} for *Phormidium* mats, an order of magnitude higher than those for epilithic communities in temperate streams. This is probably the result, not so much of high primary productivity, as of slow turnover and absence of large herbivores. The greatest loss in biomass is caused by scouring and abrasion, the mats being only weakly attached to the stream bed. Mats are absent, or confined to slow-flowing side streams or ponded areas, in rapid streams with high sediment loads. Tolerance, rather than adaptation, to the extreme conditions of Antarctic streams, seems to be the secret of their success.

Small rivers and streams occur in the coastal regions of the continent and on the peri-Antarctic islands. In the sub-Antarctic these may be permanent or seasonal, being fed from springs or carrying run-off from peaty ground. Further south, flow is unrelated to recent precipitation and occurs for only 4–10 weeks in the year, during the summer melt. Because the ambient temperature is always close to zero, flow is highly dependent on radiation balance. Streams which run over ice have already been mentioned (p. 57).

As in the Onyx River, the vegetation of streams in the maritime Antarctic is determined more by physical than chemical conditions. On James Ross Island (64°S 58°W) the bigger, fast-flowing, streams are turbid and support no visible growths. In smaller, clearer, shallower, streams perennial cyanobacterial mats grow where the substratum is stable and scour not too vigorous. Annual growths of epilithic filamentous green algae prevail in more turbid sandy situations. The green algal growths have about ten times the photosynthetic efficiency on an ash-free dry weight basis than the cyanobacterial mats, which, as in other situations (p. 57), appear to have a high proportion of inactive or senescent cells (Hawes and Brazier 1991).

5.5 Conclusions: the colonization of polar inland waters

The inland waters of the polar regions span a wide range of stability in their physical and chemical characteristics. Some of the conditions having immediate impact on organisms – nutrient availability, salinity, water movements – are little different from those in temperate regions. Low temperatures, which, however, cannot get below 0 °C in freshwaters, may slow down growth or decomposition but do not have any critical effects on tolerant organisms. The virtual absence of light for the months of the winter seems to be met by photosynthetic organisms having extremely low rates of respiration and a correlated ability to carry out net photosynthesis at very low irradiance levels, without, apparently, any other adaptations. Extreme oligotrophy is a feature of some polar lakes but is not a peculiarity of these regions. Otherwise, freshwaters in both the Arctic and Antarctic can be eutrophic with biomasses and primary productivities sometimes comparable with those elsewhere as far as can be judged from rather sparse data.

A striking feature is that the floras and faunas of these inland waters feature so few species, especially in Antarctica. This is most obvious with fish, of which the Arctic has, at the most, 11 native species and the Antarctic none, as against upwards of 40 in Britain. That fish can flourish, in sub-Antarctic freshwaters at least, is shown by successful introductions. One suspects that, similarly, there are numerous species of algae and invertebrates that could establish themselves in polar lakes if given the chance. That they have not must be put down to the relatively short time, about 10 000 years, which have been available since the ice began to retreat and colonization became possible. Land connections in the Arctic have made transport by water, wind, or birds, easier than in the Antarctic and accordingly its freshwaters have the more diverse biotas. Priddle and Heywood (1980) conclude that in the Antarctic most of the selection pressure is encountered in reaching freshwater habitats across the Southern Ocean or the ice-clad wastes of the continent. Once established, an organism encounters little competition from similar species. Consequently, even the most unusual freshwater environments have few specialist organisms and the general result is a miscellany of opportunistic species making up a rudimentary trophic structure with lakes which are physically and chemically similar frequently having different biota. The more structured communities are found at the microbial level since microorganisms are generally more easily disseminated and more resistant to the vicissitudes of passage.

The partial, or sometimes nearly complete, containment of lake systems by ice permits sustained stability of the water column and the exploitation of niches within specific depth layers by microorganisms with different metabolic capabilities. There can be tight vertical coupling between nutrient supply and demand. The dry valley lakes in particular offer a diversity of relatively simple ecosystems which can scarcely be rivalled anywhere else.

6 The marine benthic habitat

6.1 Introduction

Polar land habitats suffer extreme variation of conditions, with desiccation or mechanical damage by wind, unstable substrates, or ice movement, as the usual limiting factors. Shallow inland waters have similar disadvantages. Deeper inland waters afford more stable conditions with steady temperatures and ample water, but tend to be poor in nutrients. Benthic vegetation is frequently the most successful form of life in them. The sea similarly provides stable conditions and temperatures which cannot fall below −2 °C. This is reflected in the circumstance that a majority of polar invertebrates is *stenothermal* (i.e. unable to survive outside a narrow temperature range), and standing in contrast to the terrestrial organisms such as *Nanorchestes* (p. 83), which have remarkably wide thermal tolerances. The sea has the additional advantages that there are no solid barriers to transport of nutrients and movement of organisms. On the other hand, mechanical damage by drifting ice can be severe and polar shores usually seem barren, all exposed life down to several metres being battered and scraped off (Fig. 6.1). Icebergs may plough up the seabed down to 300 m or more. Nevertheless, benthic life is abundant in polar seas.

These considerations apply equally to Arctic and Antarctic and, of course, benthic habitats in the two regions have similar trends in irradiance and ice cover. Nevertheless, there are some striking differences:

1. Although extents of pack ice are of the same order, that in the Arctic is largely multi-year ice, covering the benthos year round, whereas in the Antarctic most of the ice melts each year. This distinction is important for abyssal benthos. Part of the Southern Ocean is covered by extensive permanent ice shelves, such as do not exist in the Arctic.

2. The Southern Ocean has open connections with the Atlantic, Pacific, and Indian Oceans, but the Arctic Ocean has only limited connection with the Atlantic and even more limited connection with the Pacific.

3. The coasts of Eurasia, western North America, and Greenland lead continuously from well below the Arctic Circle into the high Arctic whereas there is a gap of 1100 km between the most northerly point of the Antarctic continent and the nearest large land mass.

Fig. 6.1 Floating ice, Signy Island. In a rough sea such ice scours rocks in the littoral and sublittoral zones down to as much as 10 metres. (Photo G. E. Fogg.)

4. The continental shelf of Antarctica is narrow and 400–600 m deep whereas that in the Arctic Ocean is broad and at depths of 100 m, or less, to 500 m. Shallow epicontinental seas – Barents, Kara, Laptev, and Chukchi – make up nearly 36% of the Arctic Ocean area but contain only 2% of its water.

5. The Antarctic has abundant, large, tabular icebergs whereas the Arctic has fewer, small, irregular bergs, mainly in the Greenland Sea, less in the Bering Sea, and none in the Arctic basin.

6. The Antarctic seas have generally high levels of nitrate, phosphate, and silicate in the euphotic zone whereas Arctic waters have lower levels which are regularly depleted each season.

7. The sediments around Antarctica are a mosaic of muds, fine and coarse sands, and large and small boulders, all of glacial origin, whereas the Arctic has a large input of river-borne material with muds and clays predominating.

Marine benthos is not easy to study. Sampling the bottom with dredges or grabs is a hit-or-miss procedure which at the best is only semi-quantitative. However, these methods have been reasonably effective in gathering the larger species, and biologists with early expeditions made some surprisingly complete collections. Observations by underwater television provide a valuable adjunct but SCUBA diving, which was introduced into polar waters by marine biologists in the early 1960s, is an unrivalled means for observation and experiment *in situ*. For depths of more than about 60 m submersible vehicles must be used.

6.2 The littoral zone

The littoral, or intertidal, zone extends between extreme high water spring tides and extreme low water spring tides. These two limits are more theoretical than actual water levels, which vary not only with the relative motions of moon, sun, and earth, but with wind direction and atmospheric pressure. It is best to look on them as marking the centres of supralittoral and infralittoral fringes, respectively. Usually tides are semi-diurnal (i.e. with two more or less equal cycles in 24 hours 50 minutes, as on nearly all Arctic coasts. The Antarctic region has diurnal tides, with only one cycle in this period, or mixed tides intermediate between the two. The tides around South Georgia, for example, are described as irregular. Over much of the Arctic the tidal amplitude is less than 2 m, as in the Antarctic, but around the southern part of Greenland it is in the range 2–4 m. The ecologically most significant feature is that the duration of immersion in one annual cycle oscillates between 100% at the bottom and zero at the top of the littoral zone, the slowest change in duration being centred on mid-tide. To varying degrees, then, intertidal organisms alternate between the relatively stable temperature and salinity of the sea and the highly variable conditions of temperature, desiccation, and salinity encountered above sea level. It is, however, an oversimplification to regard the distribution of organisms in the littoral zone as a response to gradients in physical factors and to assume that the absence of a species from a particular level can be put down to its lack of tolerance for the conditions prevailing there. Chance plays a large part and only on a basis of study over many years of the population and community dynamics, life histories, and behaviours, of the species involved can we come to a full understanding of the zonation of organisms which is such an obvious feature of most shores.

Unfortunately, few studies in polar regions have extended over complete yearly cycles, let alone several of them. It is clear, however, that frequent storms and the overwhelming effects of ice render tidal exposure of somewhat less importance than it is elsewhere. When the sea is frozen, sea ice becomes juxtaposed to land ice but moves with the tide and is fractured into a series of parallel tide cracks. These get filled with snow and the resultant slush provides a nutrient-rich and well-illuminated medium for growth of microalgae.

6.2.1 Arctic littoral zones

Nearly all Arctic shores are ice-bound from late autumn until summer thaw. Once air temperatures have fallen below freezing, spray freezes on the shore. The littoral zone consequently supports no active life in winter and when the thaw comes it is scoured by floating ice. The only large organisms to be found in this zone at this time are those, such as red or brown seaweeds, finding shelter in crevices or under overhangs, or those which can move up from the sub-littoral. Rapidly growing microalgae, such as filamentous green forms, are able to colonize rock surfaces in the intertidal zone in the summer.

The marine Arctic has been defined as those areas in which unmodified water of polar origin is found at the surface and down to a depth of at least 200 m. The marine sub-Arctic has waters of mixed polar and non-polar origins. It is most extensive in the Atlantic sector, including the Scotian and Hudson Bay shelves, Newfoundland, the whole coast of west Greenland, the water around Iceland, and the Barents and Kara Seas off northwest Russia. These areas are still cold and occasionally have drifting ice. Spells of unusual cold can cause mortality in the littoral zone, the extent of the damage depending on how quickly and how much the temperature falls and subsequently rises, on the state of the tides, and on the prevailing weather. Species that can dry out may survive very low temperatures (p. 43) and mobile animals can take evasive action, retreating to crevices, under seaweed, pools, or the sea. Tidal pools may provide refuge but thaw water draining into them can lower salinity to the detriment of many species. Pools high up in the littoral characteristically contain green algae, which are more tolerant of salinity changes, whereas brown shore weeds occupy those in the mid-littoral zone. A succession of mild winters may allow the littoral zone to become occupied by brown seaweeds, barnacles, or mussels, all of which require several seasons to become established, then a return of cold and ice can obliterate them. The boundaries between Arctic and sub-Arctic shores may be determined, allowing for this type of season-to-season variation, by the sudden disappearance of common organisms of this type, for example, the mussel (*Mytilus edulis*), the barnacle (*Balanus balanoides*), and the periwinkle (*Littorina saxatilis* var. *groenlandica*).

Sub-Arctic littoral zones have many species in common with temperate waters in the same ocean further south. Thus Upernavik at 72°42′N on the west Greenland coast, which receives a mixture of polar water and North Atlantic water giving June temperatures of 0.2–1 °C, has brown seaweeds (*Fucus* spp.) and a comparatively rich fauna, including *Littorina*. Mussels extend from the mid-littoral into the infralittoral fringe and in the upper part of this range have large specimens of the sea anenome *Actinia* attached to them (Dayton in Smith 1990).

Fragmentary saltmarshes are found in sheltered spots on the coasts of the Canadian and American Arctic, Greenland, Iceland, northernmost Scandinavia, and Arctic Russia. Saltmarsh is a community, dominated by halophytic flowering plants, which establishes itself on stable intertidal silt and mud. It reaches its maximum development in temperate regions. In the Arctic, where intertidal deposits are frequently reworked by ice action, well-developed saltmarsh is lacking and only a mosaic of, usually, depauperate saltmarsh species is found. A grass, *Puccinellia phryganodes*, is always the primary colonist. It is not known to set seed but propagates vigorously by means of surface stolons. Accompanying it there is often *Stellaria humifusa*, *Cochlearia groenlandica*, and various sedges such as *Carex glareosa*. On the landward side saltmarsh may merge into brackish, freshwater or terrestrial vegetation. On the coast of Kotel'nyy Ostrov (75°59′N 138°00′E), where there are only 150 days in the year with mean temperatures over −10 °C, *P. phryganodes* is accompanied by half-a-dozen other flowering plants in areas which are only inundated during winter storms. In the sub-Arctic, as for example between Anchorage and the Queen Charlotte Islands, 60–54°N, there is overlap with the northern ranges of temperate saltmarsh plants such as *Salicornia*, and *Suaeda* (Chapman 1977).

The productivity of Arctic saltmarshes is low. Values based on above-ground dry matter harvested at the peak of the growing season for the dominant species are 0.02–0.24 $\mathrm{kg\,m^{-2}\,yr^{-1}}$ for *Carex ramenskii* and 0.05–0.14 $\mathrm{kg\,m^{-2}\,yr^{-1}}$ for *P. phryganodes*, both on the Alaskan coast at 69–71°N. Values for temperate saltmarsh plants are around 1 $\mathrm{kg\,m^{-2}\,yr^{-1}}$. The animal life of Arctic saltmarshes has scarcely been studied. Geese are the main herbivores (Bazely and Jefferies in Woodin and Marquiss 1997).

6.2.2 The Antarctic littoral zones

As in the Arctic, the littoral zone of exposed Antarctic coasts supports little life. On the shore of Terre Adélie, at 66°40′S 140°0′E, which is ice-covered for 10 months in the year, patches which become temporarily free of ice are colonized by algae from the sublittoral, such as benthic diatoms and the green alga *Monostroma pariotii*, but no animals have been reported as present. The lichen *Verrucaria* occurs in the supralittoral fringe and may extend lower down (Arnaud 1974).

Further north, on Signy Island, the littoral zone becomes ice-free for about half the year. In summer, the red alga *Porphyra* occurs on vertical surfaces around the extreme low water springs level. The colonizers which get highest are the green algae, *Ulothrix*, which has a particularly broad salinity tolerance, and *Urospora*, on firm substrates at about mid-tide level. Below this is a band devoid of larger algae, down to low water springs. The red seaweed, *Leptosarca*, encrusting coralline red algae, and *Monostroma* are found in crevices in the more sheltered areas. Species found in these situations contain osmoregulatory solutes and antifreezes (pp. 36 and 147; Wiencke 1996). A limpet, *Nacella concinna* (*Patinigera polaris*), is the dominant invertebrate in the shallow waters and in winter is normally confined to the sublittoral but in summer moves up into the littoral zone. There it grazes on diatoms and green algae and is itself preyed on by the dominican gull, *Larus dominicanus*, and sheathbill, *Chionis alba*. This species withstands freezing temperatures for a limited time and its return to the sublittoral is correlated with the fall of air temperature in the autumn, the operative factor perhaps being the ice film that forms on exposed rocks. Other organisms, such as coralline algae and species of spirorbid polychaete worms, grow on the limpet shells and are thus carried willy-nilly into the littoral (Heywood and Whitaker in Laws 1984). No obvious invertebrate other than this limpet is found in this zone, although crevices and the undersides of boulders may harbour a variety of animals. Where mudflats occur, both epifauna and infauna can exist intertidally.

Tide cracks, which occur around grounded icebergs as well as along the shore, have been studied on Signy Island. Seawater percolates freely through the slush of snow, which remains frozen, the winter temperature always being below zero. When snow is accumulating and seawater moving freely through it, there is abundant growth of the diatom, *Navicula glaciei*. Appearing first at the end of May, this shows vigorous growth in early September when light conditions begin to improve, and a peak is reached in early November, when the chlorophyll *a* density reaches 7.5 mg l^{-1}, with a standing crop of 5.5 mg C m^{-2}. After this there is decline until the break-up of fast ice in December (Whitaker 1977).

The littoral zones of sub-Antarctic islands, all of which are outside the impact of pack ice, have more varied floras and faunas than those just described. Heard Island (53°05′S 73°30′E), a little south of the Polar Front, has air temperatures between −10.6 and 14 °C and sea temperatures from −1.8 to 3.4 °C. Mushy ice may persist for a few hours and ice boulders are frequent on some shores but the effects of these are small. The supralittoral fringe, wetted by splash or spray but exposed mainly to aerial conditions, is characterized by black lichen but has no littorinid molluscs, as would the equivalent zone in the northern hemisphere, although terrestrial arthropods, such as mites and beetles, are present. The top of the littoral zone proper is marked by a band of encrusting coralline red algae but barnacles, which usually define this zone, are absent. The infralittoral fringe is occupied by the large brown kelp, *Durvillea antarctica*. Its fronds shelter species (e.g. a small chiton *Hemiarthrum setulosum*, the amphipod *Hyale* sp., and various small red algae), some of which are more characteristic of the sublittoral. It seems that the limpet *Nacella kerguelenensis*, which is also abundant in this zone, is prevented from invading the littoral by predation by the dominican gull. Between the coralline algae and the *Durvillea* is a zone of mixed algal species, mostly red with animals (e.g. the littorinid *Laevilittorina heardensis* and the small bivalve mollusc *Kidderia bicolor*) being found in crevices. An unexplained feature is that this zone is separated from the lichen zone above it by a few centimetres of bare rock. Despite persistently cool, damp, weather with little sunshine, some filmy algae nevertheless die through desiccation in late summer. The zones of the supralittoral, littoral, and infralittoral are not fixed absolutely in relation to tide level but, as seen elsewhere in the world, shift upwards with increasing exposure to wave action. In winter, the exposed shore is glazed by freezing sea spray. Presumably this kills off much of the summer's growth of algae (Knox 1994).

The rocky shores of Marion, Macquarie, and Kerguelen Islands are generally similar in flora, fauna, and zonation to those of Heard Island. There appears to be nothing resembling saltmarsh on any of the sub-Antarctic islands.

6.3 The shallow sublittoral zone

Only out of reach of drifting ice, around 10 m below low tide level, can sublittoral benthos develop fully in the high polar regions. Below this depth there is gouging by the keels of icebergs. Transects in shallow Arctic seas have shown as many as a 1000 of these events per kilometre with an average width of 7.5 m. During autumnal storms, gouging may overturn a muddy bottom to a depth of 30 cm. The huge icebergs of the Antarctic can scour down to depths of at least 300 m. Ice damage does not only come from above. *Anchor ice* forms on the bottom when the temperature of this is below freezing point and, being buoyant, will eventually break away, carrying organisms with it. Ice is also the cause of extreme variations in salinity. As the brine from the freezing surface layer sinks to the bottom, salinities may rise locally to 80 or 100‰ or even as high as 183‰. When the brine moves seaward it causes thermohaline circulation. On the other hand, during the summer melt and the resumption of river flow, the salinity of the shallow sublittoral may decrease.

When the sea is ice-free the penetration of light is affected mainly by turbidity, contributed by both inorganic matter in suspension and plankton. The precise evaluation of the radiation available at different depths in water is complex but for most biological purposes it suffices that irradiance falls off exponentially with depth, assuming that the water column is of uniform transparency (Fogg and Thake 1987; Dring 1992). Put simply, if irradiance is reduced by half in penetrating one metre, it will be reduced to a quarter by two metres, to an eighth by three metres, and so on: 0.1% of total photosynthetically available radiation entering at the surface may be expected to penetrate to 100 m in the clearest seawaters. The 0.1% level roughly defines the *photic zone*, in which photosynthesis is possible. In clear oceanic waters blue light penetrates most but in inshore waters, because of selective absorption and scattering by humic substances and particulate matter, the orange wavelengths have the greatest penetration. Thus, there are gradients in light quality and these can be of biological importance, as, for example, in affecting morphogenesis and reproduction. The quantum ratio of blue to red irradiation (quanta at 450 nm as a percentage of the total at 450 and 660 nm) is about 48 at the water surface but shifts to 98 and 2, respectively at the bottom of the photic zones in clear and turbid waters. Penetration of radiation is reduced by sea ice (p. 159), for example, 2 m of congelation ice (p. 158) reduces photosynthetically available radiation by about 90%, with peak transmission in the blue-green, around 500 nm. Snow cover reduces penetration still further, for example, a 70 cm thickness reduces radiation to 3% of its incident value. The sea ice algal community (p. 161) also reduces penetration of light.

The question of what is the maximum depth at which benthic algae can live is general to marine biology but it has been raised particularly with respect to polar waters. Below 40 m, growth is sparse but there are reports of macroalgae recovered from depths in excess of 100 m where irradiance is at best extremely low. For example, the green alga, *Monostroma kariotii*, has been recovered from 348 m off Possession Island (72°S 171°E). Such reports must be regarded with caution since ice may detach algae and carry them into deeper water where, because of the low temperature, they may survive in a viable state for some time. Proof must come from direct observation from a submersible. This has been done in other parts of the world and has shown that coralline red algae can live attached to the substratum at depths of 130 m or even 268 m. Calculations by Raven (1984) suggest that photosynthetic growth is just possible at a photon flux density of 1 μmol quanta $m^{-2}s^{-1}$, about the maximum which may be expected during the day at 100 m in clear water. Nevertheless, deep water red algae seem to survive at 0.05 μmol quanta $m^{-2}s^{-1}$. Their pigmentation is, indeed, such as to give maximum absorption of the blue light available at depth and it is presumed that low temperature ensures that the basal rate of respiration is minimal. Further research may show that these algae possess mechanisms making for highly efficient utilization of very low irradiance. The possibility that deep water algae in Arctic regions supplement photosynthesis by heterotrophic assimilation of dissolved organic substances has been suggested but there is as yet no direct evidence of this (Kirst and Wiencke 1995).

Within the photic zone much of the organic matter on which the benthos depends may be supplied by benthic plants. However, at greater depths the community is dependent on allochthonous organic matter which may come from plankton in the water column above, or be advected from elsewhere, from ice algae, or from debris from the land.

6.3.1 The shallow sublittoral zone in the Arctic

The Beaufort Sea, with its coast at roughly 70°N, has a shelf which is generally muddy with sandy areas nearshore and patches of gravel at the shelf break at a depth of about 70 m. These deposits are ice-borne glacial debris. The Mackenzie River and other large rivers cause seasonal fluctuations in salinity and ice covers the shelf from September through to June or July. The nearshore waters, subject to disturbance by ice and with a freshwater influence, have ephemeral populations of chironomid larvae and oligochaete worms. In deeper water, down to 20 m, there are patches of different substrata with associated communities of many species of polychaetes, bivalves, and isopods in the sediments (*infauna*), and mysids, amphipods, isopods, copepods, and euphausiids on the surface (*epifauna*). The offshore zone down to the shelf break has polychaetes making up 32–87% of the total macrobenthos, with bivalves, ophiuroids, holothurians, and many crustacea. In the relative absence of suitable rocky substrata macroalgae are not abundant (Zenkevitch 1963; Dayton in Smith 1990).

However, an isolated patch of cobbles, covering some 20 km^2, supports a stand of kelp, *Laminaria solidungula*. For eight months in the year it exists in virtual darkness under ice which is rendered almost opaque by wind-blown debris from the shore and sediment brought up by anchor ice. Surprisingly, during this period it achieves rapid growth, depleting its reserves of organic carbon in doing so. It is then ready to take full advantage of the brief summer to carry out its photosynthesis. The isolation and almost monospecific nature of this patch of vegetation gave the opportunity to use the ratio of the carbon isotopes, ^{12}C and ^{13}C, in the animals in the community, to determine the fate of the photosynthetic products. During photosynthesis there is discrimination against ^{13}C, the heavier isotope, to an extent that varies according to the type of plant and the conditions to which it is exposed. The values found in animals reflect those in the plant material they have eaten, even if it is second-hand. The Beaufort Sea observations showed more discrimination (measured as deviation, $\delta^{13}C$, from a standard and expressed in parts per thousand) in phytoplankton (i.e. around −26), than in the kelp, around −16, as compared with −7 for molecular carbon dioxide in seawater. As Fig. 6.2 shows, the herbivorous gastropod, *Margarites vorticifera*, feeds mainly on the kelp but the filter-feeding bryozoans largely on phytoplankton (Dunton and Schell 1987). The benthic primary consumers have to rely on their own reserves and detritus to carry them through the winter. In regions where kelp grows, much of this detritus consists of decaying tissue worn away from the tips of its blades. Excretion of mucilage, a feature of many brown algae, is rather slight in *L. solidungula*. Not only herbivores but some filter-feeders, such as ascidians, and even carnivores, such as the gastropod *Polinices pallidus*, come to contain high proportions of kelp carbon. The opossum shrimp (*Mysis littoralis*) incorporates much kelp carbon and since it is a major food for many vertebrates, including marine mammals, the macroalgae contribute appreciably to the higher trophic levels. Ice algae (p. 162) also contribute to benthic secondary production, 1–10% of their biomass eventually sedimenting to the bottom. However, the total primary production, including that by phytoplankton, of the Beaufort Sea is low, between 10 and 25 g C m^{-2} yr^{-1}, and rather less than the input of organic carbon in the form of peat, eroded from the shores or brought down by rivers, which amounts to around 30 g C m^{-2} yr^{-1}. Little of this peat is utilized by the macrofauna but the fact that it does not accumulate suggests that it is decomposed by bacteria. In this case, there may be input of peat carbon into higher trophic levels via meiofauna feeding on bacteria (Dayton in Smith 1990).

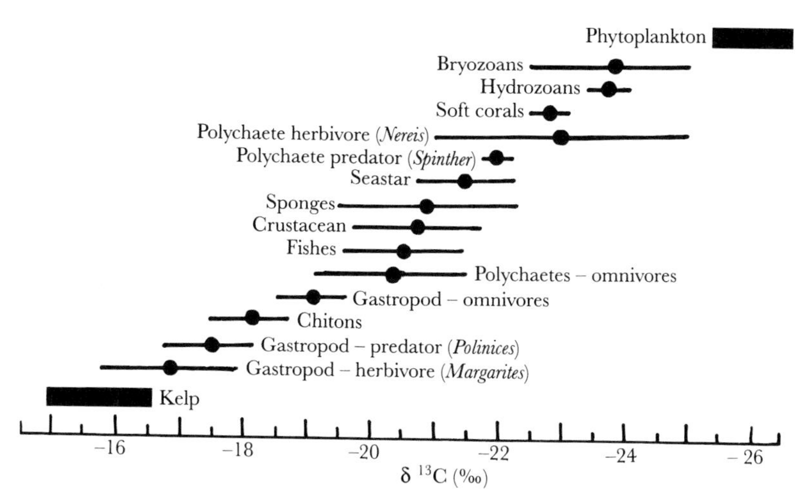

Fig. 6.2 Differences in contents of carbon isotopes, ^{12}C and ^{13}C ($\delta\ ^{13}C$), of flora and fauna from Boulder Patch, Beaufort Sea. (Redrawn from Dunton and Schell 1987, courtesy of the authors and Springer-Verlag, Berlin.)

Contrast is provided by the adjacent Bering and Chukchi Seas, which together have a continental shelf of $1.5 \times 10^6\,km^2$, stretching between 58° and 75°N and around 50 m deep. They are among the largest and most productive shelf habitats in the world. The deposits are poorly sorted mud, sand, pebbles, and cobbles. Most of the area is ice-covered during the winter but open during the summer. The benthic fauna is varied and its biomass is high. Bivalve molluscs are the most important infauna but large areas are dominated by amphipods, particularly *Ampelisca* and *Byblis* spp. There are at least 211 species of epifaunal invertebrates, most of which are molluscs, arthropods, and echinoderms. Among these, in the southeast Bering Sea, are four commercially important crabs. Species of starfish are abundant, making up some 70% of the epifaunal biomass.

Pink shrimp (*Pandalus borealis*) and crabs are important predators on smaller epifaunal species and in turn are themselves food for bigger crabs, fish, and marine mammals. Asteroids, including starfish, are generalized predators but affect bivalve populations in particular. Fin-fishes, such as flatfishes, cods, and sculpins, because of their numbers and active searching abilities, have major impact on the benthic community (Zenkevitch 1963; Dayton in Smith 1990).

The 25 or more species of marine mammals in the Bering Sea are estimated to consume between 9 and 10×10^6 tonnes of pelagic and benthic organisms per year. Most seals eat fin-fish but some, for example the bearded seal (*Erignathus barbatus*), prey on benthic invertebrates, and the walrus (*Odobenus rosmarus*), until it depleted the stock, on bivalve molluscs. The walrus is widespread over the area and, since a Soviet ban on its hunting, has multiplied considerably. It does not use its tusks for digging, as one might suppose, but supports its head on them while it extracts the soft shelled clam (*Mya truncata*) by sucking or hydraulic jetting to excavate a pit up to 30 cm deep. It may find these clams, which have conspicuous siphons, visually but if visibility is poor it furrows the upper few centimetres of sediment and identifies its prey by touch. One such furrow was found to be more than 60 m long and yielded 34 clams from depths of over 30 cm. The amount of sediment resuspended by walruses in the Chirikov Basin of the northeast Bering Sea is about 100 million tonnes per year. Pits left by walruses offer a habitat to many kinds of benthic invertebrate. A remarkable number and variety of agencies disturbing the sediments of Arctic seas can be listed. Another large mammal, excavating pits 1 × 2 m and up to 0.5 m deep is the grey whale (*Esrichtius robustus*), for which major foods are the amphipods, *Ampelisca* and *Bybles* spp. This animal is estimated to resuspend 172 million tonnes of sediment per year in the Cherikov Basin, three times the amount of sediment deposited by the Yukon River (King 1983; Ainley and DeMaster in Smith 1990).

This richness of animal life, with a mean standing stock of perhaps 300 g m^{-2}, is supported mainly by the primary productivity of phytoplankton. The hydrographic conditions in the Bering Strait are particularly favourable for algal growth during the three-month summer (p. 191). Zooplankton grazing consumes little of what is produced and much of it sediments to be used by the benthos. Benthic biomass is correlated with the primary productivity in the overlying water.

Another highly productive area is the Barents Sea. Upwards of 170 species of green, brown, and red seaweeds and 1700 invertebrates have been recorded from around its coasts and the standing stock of benthos varies from 10 to 15 up to 1000 g m^{-2} or more on the southeastern slope of the Spitsbergen Bank (*c.* 75°N 20°E). As for several other Arctic and sub-Arctic shelf regions, brittle stars are dominant in the macrobenthic fauna. The trawling industry is active, the main catches being demersal fishes, cod, and haddock. Again this productivity is based on phytoplankton, the growth of which is particularly prolific along the fronts (p. 24) lying between Svalbard and the northern Scandinavian coast (Dayton in Smith 1990).

Svalbard, 77–80°N and 10–30°E, at the northwest edge of the Barents Sea, has a rich algal flora which is associated with west Greenland and Arctic America rather than Siberia. Sublittoral algae are, however, little in evidence in some of the fjords, which because of shallow sills at their seaward ends, have heavy siltation and reduced water exchange. The fauna is also sparse, with a bivalve, *Portlandia arctica*, characteristic of the coldest part of the Arctic basin, and a polychaete, *Lumbrineris* spp., as its most prominent components. Elsewhere, the rocky seabed supports luxuriant growths of *Laminaria solidungula*, associated with the red alga *Phyllophora*, down to a depth of at least 27 m. The brown seaweed, *Scytosiphon lomentaria*, which occurs here, is of interest because its morphology is under photoperiodic control. It can develop either in the form of erect filaments under short days or as a prostrate crust under long days. A strain isolated at Helgoland, 54°N, has a critical day length between 12 and 13 hours at 15 °C. If a one-minute light break with a low irradiance of blue light is given in the middle of the dark period of 16 hours in the short-day regime, the formation of erect filaments is completely inhibited. Isolates from different latitudes show a clear relation between latitude and the temperature range in which the erect form is produced under short-day conditions. Whereas strains from 32–48°N formed erect thalli at all temperatures from 5 to 20 °C, one from Iceland, 66°N, was fully blocked in this respect above 15 °C, and one from Tromsö, 69°N, from 10 °C upwards. The Svalbard strain should be examined from this point of view (Dring 1992).

The shelf areas of the Arctic Ocean have nutrient-depleted waters and extensive permanent ice cover, which lead to low primary productivity and a paucity of benthic animal life. Sublittoral areas lack growth of macroalgae because of shortage of suitable substrates. In deeper waters, between 1000 and 2500 m, biomass is around 0.04 g m^{-2}, comparable with that in the central Pacific and much less than that found at similar depths in the Antarctic.

6.3.2 The shallow sublittoral zone in the Antarctic

Benthos goes as far south as liquid water and suitable substrata are available, that is, to at least 77°30′S in the Ross Sea, where there is ice cover 2 metres thick for 10 months in the year (Dell 1972; White in Laws 1984; Arntz *et al.* 1994). Three dominant red seaweeds show often extensive development and well-defined zonation; *Iridaea cordata* in water of around 3.5 m depth, providing that there is some protection from ice abrasion, *Phyllophora antarctica* at around 12 m, and *Leptophyllum coulmanicum* below 18 m. These zones shift downwards where thinner ice or less snow accumulation allow better light penetration. Brown seaweeds seem not to go quite as far south as the reds. In terms of biomass, benthic microalgae are more important than the macrophytes. Diatoms are abundant in the top few millimetres of sediments, accumulations of sponge spicules being particularly favourable since they afford an easily penetrable substratum with protection from grazing. Although these habitats receive only around 1% of the light incident on the sea surface, recorded biomasses in terms of chlorophyll *a* range between 47 and 960 mg m^{-2}, higher values being found in the summer than in the winter. The diatom, *Trachyneis aspersa*, at depths of 20–30 m with irradiance less than 0.6 μmol quanta $m^{-2} s^{-1}$ is shade adapted to the extent of becoming light-saturated at only 11 μmol quanta $m^{-2} s^{-1}$. Surprisingly, it is not photoinhibited at 300 μmol quanta $m^{-2} s^{-1}$ whereas other algae from the same site become inhibited above 25 μmol quanta $m^{-2} s^{-1}$. Primary productivity at the peak of development of the benthic microalgae is around 700 mg C m^{-2} day^{-1} – about the same as that for the phytoplankton. Information on the annual benthic production is lacking but it seems to make an important contribution to the total in the area (Knox 1994).

The epifaunal benthos of McMurdo Sound shows three distinct vertical zones (Fig. 6.3). The top 0–15 m is a bare zone with a substratum of rock,

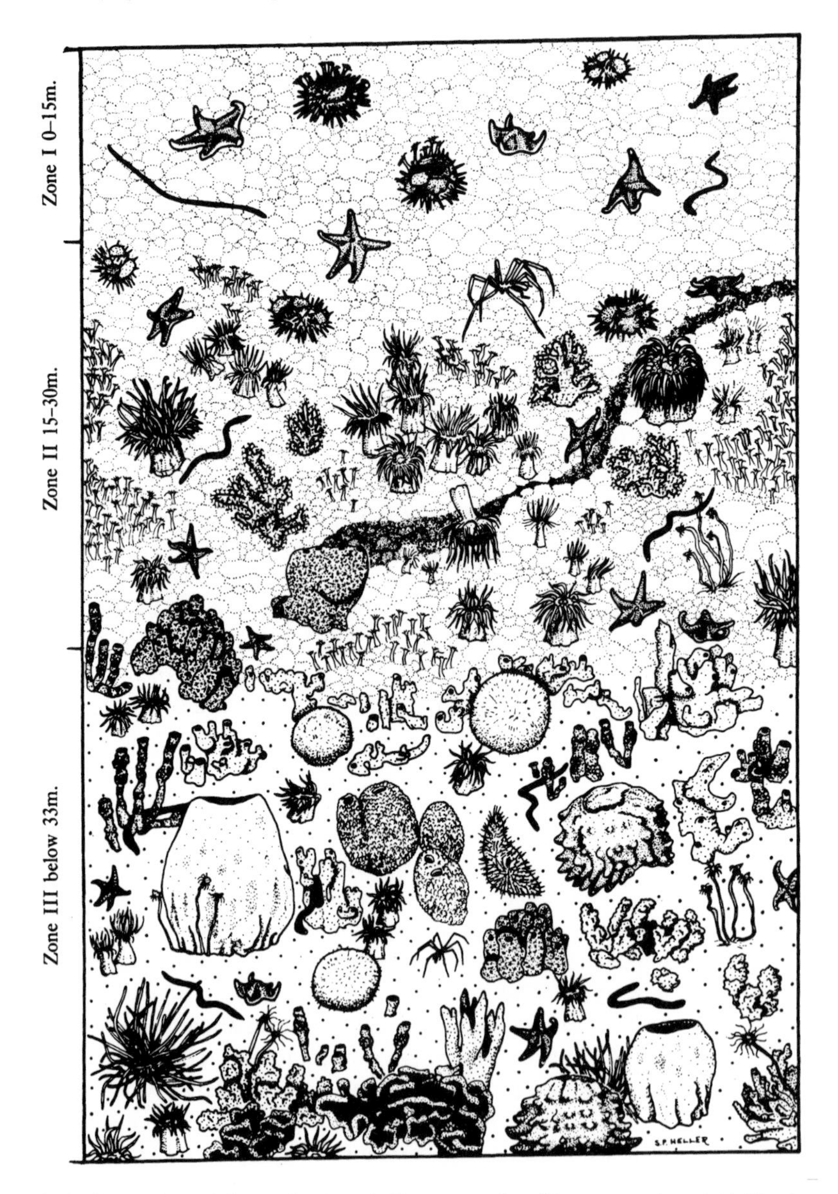

Fig. 6.3 Vertical zonation of fauna in the shallow water benthic community of McMurdo Sound. A few mobile animals, but no sessile forms, are found in Zone I; the sessile animals in Zone II are mostly coelenterates and those in Zone III are predominantly sponges. (From Dayton *et al.* 1970, courtesy of the authors and Academic Press, London.)

pebbles, and volcanic debris, devoid of sessile animals because of ice scour and disruption by anchor ice. It is briefly invaded during the summer by mobile animals including a detritus-feeding echinoid, a starfish, a necrophagous nemertine, isopods, occasional pycnogonids, and fish. The zone below has a cobble bottom with coarse sediment in between. It is inhabited by abundant soft corals, anemones, hydroids, and ascidians. The most conspicuous among the many different forms are sessile coelenterates, the alcyonarian *Alcyonium paessleri*, and anemones such as *Artemidactis victrix*, and *Hormathia lacunifera*. The mobile animals found on the bare zone are also found here, notably the fishes, *Pagothenia bernacchii* and *Trematomus pennellii*. Since the freezing point of the body fluids of teleost fish is normally above that of seawater – unlike invertebrates which are iso-osmotic or slightly hyperosmotic relative to seawater – this habitat is hazardous because the ambient water is near its freezing point and contact with ice is difficult to avoid. These benthic ice-foraging fish could not exist without antifreeze glycopeptides in their various fluid compartments (p. 37). They remain mainly inactive on the bottom, hidden in crevices to avoid seals, or perched on sponges, thereby getting a better view of the water column. Anchor ice wreaks havoc in this zone when it encases and tears off masses of plant and animal life weighing up to 25 kg. This plays an important part in determining zonation (Dayton *et al.* 1970). It may be noted that the trawls and dredges of marine biologists can do similar damage. The third zone, normally free of anchor ice, begins rather sharply at 33 m and continues down to at least 180 m. Sponges of great variety of form cover up to 55% of the ground. The abundance of these gives a unique character to Antarctic benthos and provides cover, ecological niches, and food for a great variety of other animals. Glassy (siliceous) sponges, in other parts of the world confined to deep water, are particularly abundant, perhaps because oceanic water extends right to the edge of the Antarctic continent (Dell 1972). Among the most prominent are *Rosella nuda* and *Scolymastra joubini*, both known as white volcano sponges, which are up to 2 m tall and 1.5 m diameter (Fig. 6.4). Apart from these there are anemones, the alcyonarian already encountered in the zone above, hydroids, polychaetes, bryozoans, ascidians, and many molluscs. The total biomass of this sponge community is around 3 kg m^{-2}. Spicules from the sponges form a dense mat, varying in thickness from a few centimetres to more than 2 metres, which provides a habitat for an abundant infauna. 12 500 individuals of an unspecified number of species, belonging to the crustacea, polychaeta and other groups of worms, were counted in 1 dm^3 of this material.

The dynamics of these communities are complex. Sponges, with the exception of *Myacale acerata*, which increases its mass by as much as 67%, grow at rates which are imperceptible in the course of one year. *Myacale* has an advantage in competition for space and, indeed, sometimes overgrows other species, but this is offset by heavy selective predation. Sponges live by filtering out particulate organic matter and, probably, the concentration of

Fig. 6.4 D. G. Lillie with siliceous sponges (the one he is holding was probably *Rosella villosa*) from the Ross Sea; *Terra Nova* expedition 1911–13. (From Huxley, 1913 *Scott's last expedition*, Smith, Elder & Co., London.)

phytoplankton in the southern Ross Sea is high enough to provide for them adequately. A variety of predators, including starfish and the nudibranch *Austrodoris mcmurdensis*, feed on sponges, which nevertheless maintain large standing crops. The fish are generalist feeders, taking, among other things, polychaetes, fish eggs and small fishes, and amphipods. In contrast to the situation in the Bering Sea, none of the marine mammals which frequent the Ross Sea feed on benthic infauna. The infauna in McMurdo Sound, comprising both deposit and suspension feeders, perhaps owes its luxuriance to the absence of disruption from this quarter. Another factor is certainly the high productivity of the benthic microflora. Differences between the infaunal biomasses on the east and west sides of the Sound are correlated with microalgal production. This in turn is related to currents, the east coast receiving water from the open sea whereas the west coast gets deoxygenated water flowing out from under the Ross Ice Shelf (Knox 1994; Arntz *et al.* 1994).

It is difficult to make precise comparison s between benthos in different parts of the high Antarctic. Sampling methods have varied and the taxonomic basis still leaves much to be desired; identifications have mostly been made with pickled specimens by taxonomists who have not seen the living organisms, and there is a substantial backlog of undescribed collections. As far as one can tell, sublittoral communities all round the continent broadly resemble that just described. On the coast of Terre Adélie, almost on the Antarctic Circle at 66°33′S, the assemblages of organisms are generally similar to those in McMurdo Sound but brown seaweeds, such as *Phyllogigas* (*Himantothallus*) *grandifolius* and *Desmarestia menziesii*, are more in evidence and the diversity of fauna is greater (Arnaud 1974). Around the islands a tendency of the flora and fauna to become richer in species as one goes northwards is maintained. After the poverty in species of the terrestrial and freshwater habitats of Antarctica the richness of the marine benthos is astonishing. Roughly 100 species of macroalgae and over 3000 species of the more conspicuous kinds of benthic invertebrates have been recorded from the Southern Ocean (Dell 1972, Arntz *et al.* 1994).

For benthos under moderately severe conditions which do not usually include formation of anchor ice (see Dell 1972; Arnaud 1974; Heywood and Whitaker in Laws 1984; White in Laws 1984; Dayton in Smith 1990; Knox 1994, Arntz *et al.* 1994) we may consider first that of the shallow sublittoral zone of Signy Island. Substrates vary from sand–silt, through gravel, to pebbles and cobbles. The ice abrasion zone, extending down to 2 m, bears an impoverished flora appearing only in the fast ice-free period and consisting of diatoms and the brown alga *Adenocystis utricularis*. Below this, from 2 to 8 m, growing on boulders are brown seaweeds, *Desmarestia* spp., and an underflora of red weeds. *Desmarestia* may be up to 1 m long and form dense stands with a cover of about 75% on boulders, falling to 22% on gravel. The mean biomass is in the range 230–830 g wet weight m^{-2}. The next zone down, 8–11 m, is dominated by another brown seaweed, *Himanthothallus grandifolius*, which attains a cover of about 39% on boulders and 33.5% on gravel with biomasses of 150 and 40 g fresh weight m^{-2}, respectively. These standing crops are less than those of comparable growths off Anvers Island (64°S 64°W), which range from 1.64 to 6.34 kg m^{-2} – much the same as in temperate waters. Off King George Island (62°14′S 58°41′W), *Desmarestia* spp. are the most successful seaweeds under stable conditions but are replaced by *Himanthothallus* where substrates are exposed to turbulence or impact of icebergs. Below 11 m, around Signy, is a diminished flora of *Himanthothallus*, *Desmarestia*, and red algae.

The sublittoral macroalgae can survive dark periods for as long as a year and grow at low irradiance levels, light saturation occurring between 4 and 20 μmol quanta $m^{-2} s^{-1}$. They become photoinhibited under high irradiance levels in summer but have a photoprotective mechanism which dissipates excess energy as heat. Growth occurs only at temperatures below 10 °C (5 °C for *Desmarestia* spp.) and upper survival levels are usually less than 17 °C (Wiencke 1996). Growth in some of these algae begins before the ice breaks up and, as with *Laminaria solidingula* in the Arctic, is probably triggered by day length. These are 'season anticipators' in contrast to the opportunist 'season responders' characteristic of the littoral zone (Kain 1989). However, a red seaweed, *Myriogramme mangini*, common in the Signy sublittoral, appears to be a season responder, not growing in winter. Detailed studies *in situ* on whole plants showed that photosynthesis was saturated between 10.5 and 18 μmol quanta $m^{-2} s^{-1}$ in early spring and autumn respectively, with compensation occurring from 2.5 to 2.8 μmol quanta $m^{-2} s^{-1}$. From the photosynthesis measurements and records of solar radiation and light attenuation in the water over the year, annual net production can be estimated as 1.5 and 0.007 g C g dry $weight^{-1} yr^{-1}$ at 5 m and 20 m depth, respectively. From this a maximum depth of occurrence of this species can be predicted as 22.9 m, at 2.7% of surface radiation. That it is, in fact, only occasionally found below 14 m may be attributed to competition from other algae (Brouwer 1996).

Beds of seaweed provide shelter for various fish, *Notothenia coriiceps*, *N. gibberifrons*, and *Trematomus newnesi*, together with molluscs and isopods. There are also annelids and nemerteans. Exposed rock supports a variety of echinoderms, pycnogonids, tunicates, sipunculids, sponges, hydroids, and bryozoa. The biomass of this fauna ranges from 4200 g wet weight m^{-2} on a rock overhang with mainly filter-feeding animals, to between 218 and 1723 g wet weight m^{-2} amongst the macroalgae. Estimates of the biomass of infaunal communities in mobile substrates, in which bivalves are dominant, range between 307 and 789 g wet weight m^{-2}. The large bivalve, *Laternula elliptica*, could not be dealt with by the sampling method used, but if it were included these values might increase up to as much as 2600 g wet weight m^{-2}.

Many of these benthic animals are suspension feeders and accordingly feed most actively during the summer peak of microphytoplankton. They have been presumed to cease feeding during winter but a study of bryozoans off Signy Island has shown that periods of near-zero activity during winter are actually short or non-existent. At this time they evidently feed on ultraplankton (Clarke and Leakey 1996).

Around the sub-Antarctic islands, in the absence of ice abrasion, really large kelps make their appearance. *Macrocystis pyrifera*, which may be up to 40 m in length and has fronds with gas bladders which float at the sea surface, is the most spectacular. Together with other large seaweeds, such as *Durvillea antarctica* and *Himanthothallus*, it is found around the coasts of South Georgia and Kerguelen. *Macrocystis*, especially, has an important effect on the littoral and sublittoral zones in general since it acts as a natural breakwater and reduces wave action (Fig. 6.5). Additionally, the holdfasts and fronds of these kelps provide habitats for a rich variety of smaller algae and animals. Large numbers of the delicious small bivalves of the genus *Gaimardia* occur

Fig. 6.5 King Edward Cove, South Georgia. The bed of giant kelp, *Macrocystis*, just offshore acts as a natural breakwater. (Photo G. E. Fogg.)

attached to *Macrocystis*. Kelp beds are a favourite habitat for fish, where they prey primarily on crustacea, fish larvae, polychaetes, and molluscs. One species, *Notothenia neglecta*, grazes actively on the kelp itself. All this wealth of life provides food for large numbers of seabirds.

Some general characteristics of Antarctic benthic invertebrates should be mentioned. Slow, seasonal growth and delayed maturation are features which they share with Arctic species. Food seems to be the main regulatory factor here, rather than low temperature. Slow growth and delayed maturation result in large size in some Antarctic invertebrates; giant pycnogonids, isopods, amphipods, sponges (Fig. 6.4), and free-living nematodes are commonly reported and critical examination shows that the phenomenon of gigantism has some reality. However, most polar species are smaller than their counterparts in temperate regions, especially those that deposit calcium carbonate in exoskeletons or other structures, presumably because the solubility of calcite increases with with decrease in temperature so that its deposition becomes limiting. In temperate and tropical seas, benthic invertebrates typically have pelagic larval stages but it has been supposed that in polar regions few species do this, adopting instead some form of brood protection (Dell 1972). Large yolky eggs enclosed in capsules which are brooded are common in both Arctic and Antarctic molluscs. Eighty per cent of Antarctic sponges are either viviparous or show brood protection as against about 55% of those in tropical or temperate waters. Invertebrate groups that characteristically have pelagic larvae and which are dominant in the littoral and sublittoral elsewhere (e.g. prawns, lobsters, crabs, and barnacles), are poorly represented in polar regions. On the other hand, other crustacean groups, the amphipoda and isopoda, which can be regarded as pre-adapted to high latitudes in this respect, are abundant. The relative absence of pelagic larval stages has various explanations. It avoids the hazard of surface waters where salinity may fall abruptly following ice melt and of turbulent seas with strong currents which might carry larvae away from suitable habitats. By the same token, of course, it limits dispersal and colonization. With the period of high phytoplankton production being so short it is also presumably of advantage to liberate juveniles in a stage of development at which they are best able to profit from the brief abundance of food. The recognition that ultraplankton production continues during the winter (p. 191) takes some of the force from this argument. Consideration of the total energy requirements of reproduction leads to the conclusion that production of larvae from large yolky eggs is the more efficient under conditions of poor food supply and low temperature. Nevertheless, it appears that past surveys may not have been complete and that larvae of echinoderms, annelids, and nemerteans may sometimes occur in the plankton. There is evidence that these obtain particulate food from the water column – again it may be ultraplankton – especially near sea ice, during the winter. The 'rule' that protected development is more common towards the poles thus seems open to question (Knox 1994; Shreeve and Peck 1995).

6.4 The benthos of deep Antarctic waters

True deep sea (abyssal) benthos has received little attention. The deep shelf mixed assemblage, found in the Ross Sea down to 523 m on fine sediments with erratic boulders, includes substantial populations of polychaetes, bryozoans, gorgonian corals, ophiuroid starfish, and crinoids. Other types of sediment have their own characteristic faunas (Dayton in Smith 1990). On the southwestern Weddell Sea shelf biomass wet weights up to 1.6 kg m^{-2} were found around 250 m depth but at 2000 m biomass had dwindled to less than 1 g m^{-2} (Arntz *et al.* 1994).

6.5 Benthos under ice shelves

Ice shelves, which scarcely exist in the Arctic, are enormous in the Antarctic, covering a total of more than 1 400 000 km^2 or 7% of the total ice-covered area. The Ross Ice Shelf (530 000 km^2) and the Filchner–Ronne Ice Shelf in the Weddell Sea (400 000 km^2) are the largest. They vary in thickness from about 200 m at the seaward, floating, edge to about 600 m where they join the inland ice sheet. That life can exist, at least for a limited time, under these shelves is shown by the presence of marine fish in epishelf lakes (p. 127). The question is whether resident communities are present. They do certainly exist near the seaward edge of shelves. Near White Island and Black Island (*c.* 78°S 166°E), which emerge from the Ross Ice Shelf in

McMurdo Sound, there are dense populations of filter-feeders living in darkness 30–50 km from the seaward edge (Knox 1994).

Sampling through a hole drilled in the Ross Ice Shelf more than 400 km from the open sea confirmed the presence of living organisms but evidence that these were permanent residents below the shelf, rather than strays, is equivocal. Some crustacea were found in the water column but the sediments seem devoid of infauna or sessile epifauna. This barrenness was confirmed by underwater television. Microbial biomass, measured in terms of adenosine triphosphate (ATP), was highest at the sea floor, where it was around 1 mg of bacterial carbon m^{-2}, much lower than found in continental shelf sediments and similar to that reported from impoverished abyssal regions (Knox 1994). One cannot take results from one sampling point as representative but, considering possible sources of organic carbon for support of undershelf life, one is not encouraged to think that there can be much more life than that already found. Photosynthesis is obviously impossible. Chemosynthesis may occur. There is volcanic activity in the vicinity of the Ross Ice Shelf so that seepage of water carrying hydrogen sulphide might sustain populations based on sulphide-oxidizing bacteria. However, hydrothermal vents such as provide for this type of community in mid-ocean are unlikely to be present under this ice shelf since they would produce obvious thinning of the ice. Another possibility is that seepages of methane, which has been detected in Antarctic marine sediments, might support methane-oxidizing bacteria as primary producers. Methylotrophs have been isolated from Antarctic soils but not yet from sediments under ice shelves. The probability is that such production as occurs is based on heterotrophic assimilation of dissolved organic matter. Seawater contains around 5 mg l^{-1} – or maybe more, there is disagreement about analytical methods – of organic matter, most of which is refactory high molecular weight polymeric material. However, some bacteria may be capable of breaking it down slowly and there is a proportion of more labile material. Concentrated by adsorption on sediment particles it might well provide support for a minimal community. Water emerging from under the shelf does seem to be depleted in dissolved organic matter. The turnover time might be between 1 and 10 years, the concentration being maintained by slow exchange of water with the open sea (Karl in Friedmann 1993).

6.6 Comparisons and conclusions

Below the reach of floe ice, the sea bottom provides a stable environment in which desiccation, the major limiting factor for life in polar land habitats, is inoperative. Light is a limiting factor for the plants but in shelf waters the benthos also receives organic material produced by photosynthesis in the water column. The biomass per unit area in the polar benthos is usually greater, often much greater, than that on nearby terrestrial sites, although rather less than that of benthos in temperate or tropical regions.

Predictable environments with an intermediate level of stability generally have greater species diversity than unstable ones and this is evident in the polar benthos. Diversity is also promoted by the variety of substrates available. This high diversity is coupled with a tendency to even distribution in the abundance of the different species. In one area of McMurdo Sound, 11 species were found to have population densities of more than $2000\,m^{-2}$. Historically, the surprising diversity of polar benthos may be attributable to the circumstances that expansion of the ice caps would not necessarily obliterate benthic biota as completely as those on land, that transport of propagules by water movements is less hazardous than by air, and that the more uniform conditions of the seabed are more favourable to establishment than those on land (p. 49).

There remain to be explained striking differences in the composition and diversity of the benthos in the two areas. There are a few bipolar species. Some species of the seaweeds, *Ulva*, *Enteromorpha*, and *Ceramium*, are cosmopolitan in distribution and the green alga, *Acrosiphonia arcta*, and the brown seaweed, *Desmarestia viridis*, have disjunct bipolar distributions (Fig. 6.6). The genetic affinities of these disjunct populations have been established by sequencing of nucleotide in ribosomal DNA. Comparing northern and southern samples, *Desmarestia* showed only one base change among 1073 nucleotide positions and *Acrosiphonia* exhibited 17 variable sites among 626 nucleotides. This indicates that for both species the disjunction has been recent. The temperature requirements of strains of *Acrosiphonia* show that those from polar habitats have growth optima between 0 and 10 °C and upper survival temperatures of around 22 °C whereas for cold–temperate

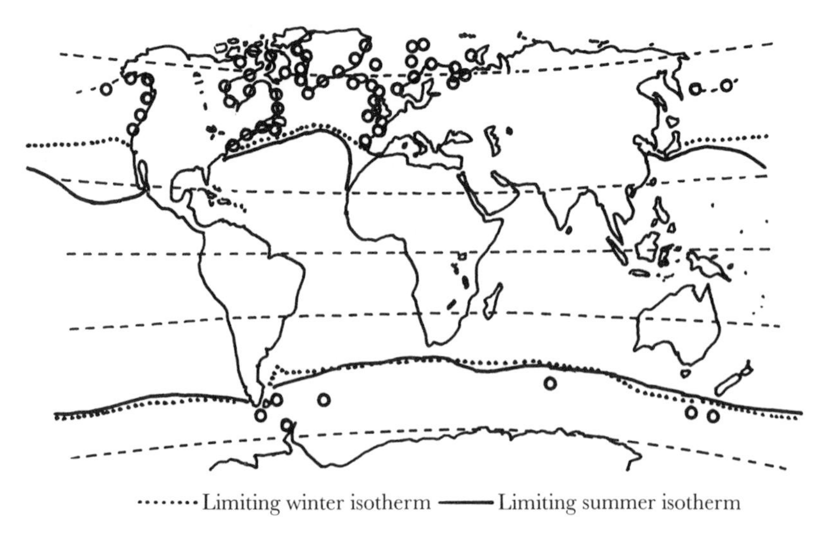

Fig. 6.6 Disjunct bipolar distribution of the green alga *Acrosiphonia arcta*. (Redrawn from Bischoff and Wiencke 1995, courtesy of the authors and the editor, *European Journal of Phycology*.)

strains these temperatures are respectively 15 °C and 23–25 °C. The upper survival temperatures suggest a possibility of transfer across the equator during the Pleistocene lowering of seawater temperature in the tropics, 18 000 BP. Growth would not have been possible during this passage. The presence of *Acrosiphonia* on the east coast of Chile and its absence from the Atlantic coasts of South America and South Africa indicate that the route of transfer may have been through the East Pacific, where at present, the tropical belt is relatively narrow (Bischof and Wiencke 1995).

A contrasting situation is presented by priapulid worms, relicts of a phylum which evolved in the early Cambrian. The bipolar disjunct pairs of species *Priapulus caudatus/P. tuberculatospinosus* and *Priapulopsis bicaudatus/P. australis* are morphologically similar but have genetically distinct enzyme patterns. In the long period since the populations were separated, evolution at the molecular level has been faster than that in anatomy (Schreiber *et al.* 1996).

Apart from the few cosmopolitan and bipolar species, the floras and faunas are quite different. Although individual numbers are questionable there is a consistent tendency for the numbers of species in the various groups of animals to be greater by a factor of between 1.5 and 6 in the Antarctic than in the Arctic (White in Laws 1984). The corresponding data for algae are ambiguous. Another feature is that whereas Arctic species show little endemism and have close affinities with cold–temperate Atlantic or Pacific forms, a high proportion, ranging between 57 and 95%, of endemic species occurs in Antarctic animal groups. Endemism at the generic level is less, between 5 and 70%. Despite its richness in species the Antarctic fauna is marked by the absence of certain major groups; many forms common in the Arctic (e.g. crabs, flatfishes, and balanomorph barnacles) are not represented in the Antarctic.

There are differences between the present-day benthic habitats in the Arctic and Antarctic. The Antarctic, lacking appreciable freshwater run-off from the land and the massive bioturbation seen in the Arctic, is the more stable and this has been favourable to diversification. The Arctic benthos is more accessible to invading species whereas the Polar Front and lack of shallow sea approaches are impediments for forms migrating into the Antarctic. However, it is necessary to look to the past for the full explanation of the differences in benthic floras and faunas. Those in the Arctic are mainly the outcome of events in the Quarternary period, that is, the last 2 million years. An Arctic estuarine water mass was formed by the inputs from the Siberian rivers and was colonized by a mixture of freshwater and euryhaline immigrants which distributed themselves around the polar basin during the last sea level transgression. These assemblages seem, however, to have been nearly eradicated during the last major glaciation. In post-glacial times there has been invasion by low temperature-tolerant species from the Atlantic and, to a lesser extent, from the Pacific. Thus, the present flora and fauna comprise relatively young assemblages with notably few endemics. An odd thing is that, whereas animal species with Pacific affinities are numerous in the inshore regions between 145°E and 120°W, algal species with the same origin are scarce. This might arise from the greater vulnerability of seaweeds requiring hard rock and shallow water (Dunton 1992). The highest diversity occurs in areas of hydrographic mixing. Areas of lowest species-richness in the Arctic benthos are those influenced by brackish waters and the deep sea. The benthos is subject to much the same range of low temperatures in the Arctic and Antarctic but nevertheless there are differences in the degree to which organisms are adapted to this. Endemic brown algae of the genus *Desmarestia* in the Antarctic have growth optima around 0 °C, an upper limit for growth at around 5 °C, and a survival limit at 13 °C. In the Arctic, the endemic *Laminaria solidungula* grows up to 16 °C and survives at 18 °C. A few Arctic species have temperature requirements similar to those of Antarctic species but on the whole the situation is consonant with the idea that the southern species have had much longer to adapt to the polar environment. The Antarctic became cut off by the Circumantarctic Current about 25 million years ago and then began to cool, although for a time it continued to enjoy a mild climate (p. 28). There is fossil evidence of a focus for the appearance of cool water taxa at high southern latitudes around this time. It seems that much of the present flora and fauna of Antarctic waters is descended from forms which originated then. This would explain the large numbers of endemics, a tendency which would be reinforced by the hydrological barriers and the propensity of the dominant benthic invertebrate groups not to have pelagic larval stages. In addition to the relict autochthonous flora and fauna there are some species to be distinguished (e.g. among the sponges), which appear to have emerged from deep water into shallower coastal waters. There are also cool–temperate species, mostly from South America, which have invaded via the chain of islands in the Scotia Arc. This has involved negotiating numerous gaps and pitfalls that eliminate whole families as well as individual species. Apart from those that appear to have followed this path, most Antarctic species have circumpolar distributions (Dell 1972; Dayton in Smith 1990).

The longer time which has been available for evolution of polar biota in the Antarctic as compared with the Arctic is evident in the physiology of organisms as well as in their taxonomic features. Endemic brown seaweeds, for example, have lower requirements for light and their growth is inhibited by rise in temperature to a greater extent than it is in corresponding endemic Arctic species (Wiencke 1996). In neither polar region has any specific mechanism for withstanding the frigid environment been evolved.

7 Sea ice as a habitat and ecological interface

7.1 The ecological roles of sea ice

The profound effects of sea ice on the climatic and oceanographic environments in which polar organisms live have been outlined in Chapter 1. It is more variable and heterogeneous in structure than land ice and since it is in contact with liquid water and consequently does not reach such low temperatures as does land ice, it provides a variety of niches for active life. Sea ice has a unique ecological role. It interposes a solid interface between two fluid phases – biologically productive seawater and air, unproductive but allowing exceptional mobility to animals able to take advantage of it. It modifies environmental conditions in the sea below and also provides a platform on which air-breathing birds and mammals can live, breed, and base foraging forays into the water. Although its annual cycle of advance and retreat is a recurrent pattern, the distribution and local structure is irregular. Its variations from year to year affect all levels in the marine ecosystem.

The ice edge has attracted attention as one of the ecologically most interesting regions of the polar oceans. Terrestrial ecologists have long been aware of the special nature of transitions between two or more diverse communities, such as forest or grassland – so-called *tension belts* or *ecotones*. These communities usually contain many of the organisms from the neighbouring ones as well as those characteristic of, and sometimes restricted to, the ecotone. Ecotones commonly have greater species diversity and greater biomass density than do the flanking communities, evidently because the variety of niches is greater and organisms may be able to draw on the resources of the adjacent environments. The ice edge is an ecotone and, like the terrestrial examples, shows increased species diversity and greater productivity. The physical and chemical mechanisms underlying this 'edge effect' are, of course, different and, unlike ecotones on land, the ice edge has no fixed location.

7.2 The physical characteristics of sea ice

The formation of sea ice begins with *frazil ice*, consisting of ice crystals, minute platelets or needles at first but growing to as much as 2–15 cm across and 2 mm thick (Gow and Tucker in Smith 1990). Continued freezing and clumping of crystals produces a slushy mixture called *grease ice*. Under calm conditions the frazil crystals freeze together to form a solid continuous cover, between 1 and 10 cm thick. Usually, however, wind-mixing prevents this and frazil is driven downwind to accumulate in thicknesses of up to 1 m in the convergences between wave circulations. Sustained wave action moulds frazil into *pancake ice*, circular masses up to 3 m in diameter, of half-consolidated crystals, which come to have upturned rims through constant bumping against each other. Eventually the pancakes join together to form a continuous sheet. Thereafter, crystal growth can only be in a vertical direction so that the ice now assumes a columnar structure. This is *congelation ice*. It has plates of pure ice with spaces containing brine in between them. This is in marked contrast to frazil ice, in which there is a branched and interconnecting system of brine channels without any predominantly vertical orientation. The structure of channels in either type of ice may be studied by making casts using water-soluble resin which can be polymerized by UV irradiation at −12 °C (Fig. 7.1).

With daily variations in surface air temperature, the concentration and distribution of the trapped brine alter, often abruptly. Warm conditions

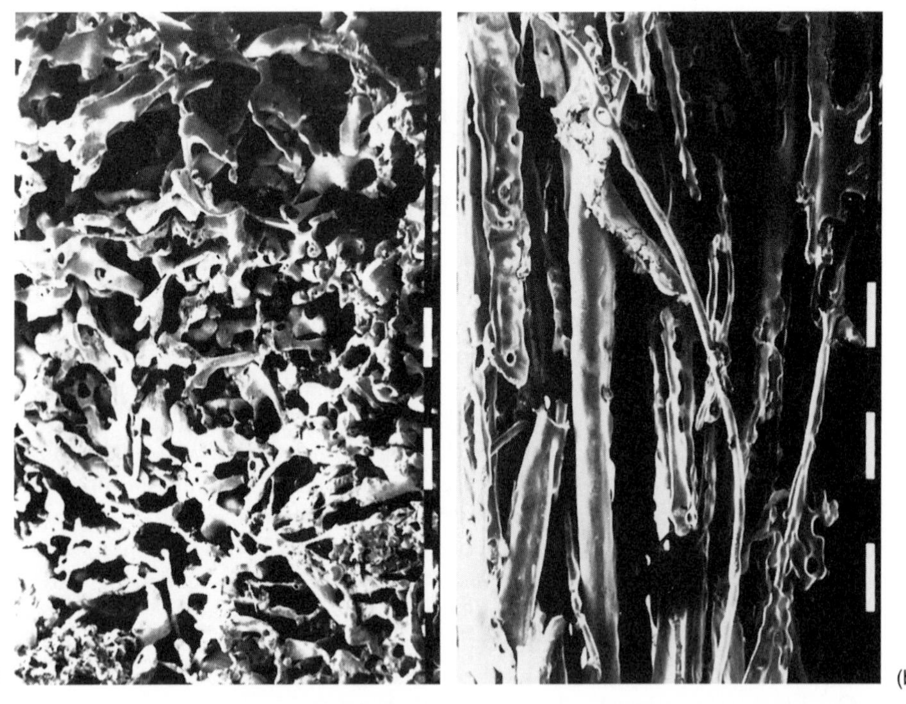

Fig. 7.1 Casts of brine channels in (a) frazil ice, composed of small granular crystals, (b) congelation ice, composed of elongated, columnar, crystals. Scale units = 1 mm. (From Weissenberger *et al.* 1992, courtesy of the authors and the editor, *Limnology and Oceanography*.)

promote the coalescence of originally unconnected channels so that eventually the brine can drain away. When large pockets of brine empty an ice stalactite may be produced. In addition to formation of new ice on its underside, the sheet is added to by snow accumulating on top and by freezing of any interstitial melt water or seawater. Snow ice is coarser grained and bubbly compared with that formed from frazil. However, ablation from the top of the ice also goes on and tends to balance these additions so that sea ice is a homeostatic system, maintaining a thickness averaging about 3 m (Melnikov 1997).

A schematic representation of the structure of sea ice in relation to the biota it supports is given in Fig. 7.2. There are three principal communities associated with sea ice, one within the ice itself, one on and in the underside of the ice, and thirdly, that comprising the warm-blooded animals that live on its surface.

Sea ice is generally less transparent to radiation than clear lake ice. Peak transmission is in the blue-green, around 500 nm, but this is shifted if algae are present towards the yellow-orange, around 600 nm. Both the intensity and quality of irradiance varies from place to place in relation to patchiness in the ice, its surface configuration, and proximity to open water channels. Whereas the temperature at the bottom of the ice is near the freezing point of seawater, −1.9 °C, that near the surface is approximately that of the mean air temperature and may be as low as −10 °C.

Apart from the logistic difficulties of *in situ* study of sea ice communities there are problems in sampling. One arises from the heterogeneity of the ice. Another is that much of the sea ice biota lives in the brine phase and

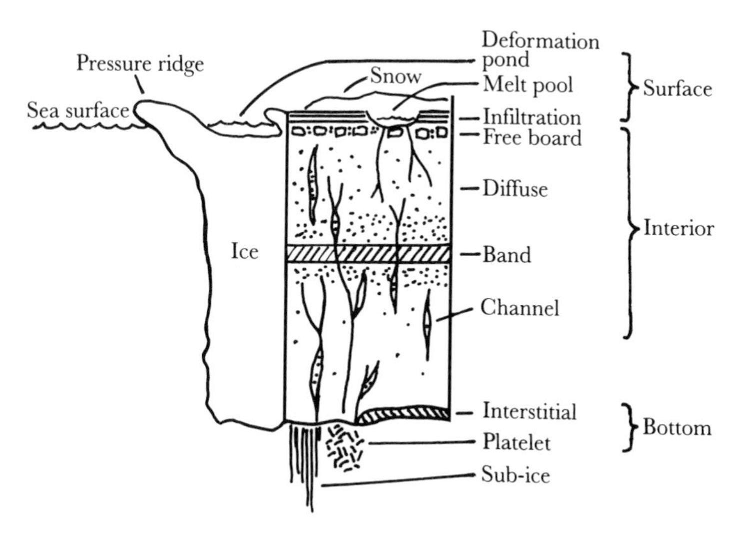

Fig. 7.2 Schematic representation of habitats and biological communities in sea ice. (Redrawn from Horner *et al.* 1992, courtesy of the authors and Springer-Verlag, Berlin.)

when a sample is melted for examination the organisms are subjected to severe osmotic shock. In fact, if samples are compared which are treated in this way with pore water samples collected without allowing the ice to melt, there is a considerable difference between the two. Dinoflagellates, other autotrophic and heterotrophic flagellates, and ciliates are particularly susceptible to osmotic shock, with losses amounting to as much as 70%, whereas diatoms are little affected. The effect may be minimized by allowing samples to melt in larger volumes of seawater to buffer salinity and osmotic changes. Not all studies have included such precautions and thus their results must be treated with caution (Garrison and Buck 1986).

Constant motion caused by wind, currents, and wave action breaks up ice and alters its morphology. The enormous pressures that can be generated by wind blowing across an ice field produce ridges, hummocks, and rafting – in which floes override each other. *Pack ice* – the term denotes any area of floating ice that is not *fast* to land – may be *open*, with plenty of water visible, or *close*, with channels or *leads* between large areas of continuous ice (Fig. 7.3). Leads are ten to hundreds of metres wide and may be several kilometres long.

Polynyas are large enclosed openings in the ice cover, tending to occur in fixed positions (Muench in Smith 1990). They arise because of upwelling of warmer water (*sensible-heat polynyas*), as with the huge Weddell Polynya in the Antarctic, or by divergence of pack ice away from a coast (*latent-heat polynyas*), as with those in the Bering Sea or off the west coast of Greenland. Latent-heat polynyas are maintained by continued removal of ice by wind,

Fig. 7.3 Pack ice with leads and iceberg, Weddell Sea *c.* 64°S, January. (Photo G. E. Fogg.)

a windward shore preventing its replacement from elsewhere. The latent heat released by ice formation is dissipated into the atmosphere and the brine-produced sinks, thereby modifying the circulation of the adjacent sea area. These polynyas are a prolific source of ice, for example, it is estimated that the Bering Sea ice cover is regenerated between two and six times from this source in one season. Polynyas are of oceanographic importance as pathways for heat losses to the atmosphere. Biologically, they provide open water in winter for birds (p. 171) and sea mammals and present an ice edge having enhanced productivity (p. 175). Polynyas appear to be more important in this way in the Arctic than they are in the Antarctic.

It is surprising that although Arctic and Antarctic sea ices are subject to the same physical laws in their formation and ablation, there are significant differences between them. These arise because the Arctic Ocean is nearly landlocked whereas Antarctic ice extends over an open sea on which cold katabatic winds, blowing off the continent, can rapidly thicken inshore ice but at the seaward ice margin warmer winds, warmer water, and wave action work together to break the ice into small floes. The unconstrained nature of the Antarctic ice edge allows wave action to extend further into the pack than it does in the Arctic. Antarctic sea ice has few melt pools, evidently because the drier air and higher wind speeds increase heat losses to the atmosphere so that surface air temperatures must be well above zero for melting to begin. In contrast, Arctic ice melts at the surface in summer to the extent that 60% of the surface may be covered by pools. The lowered albedo which results allows more short-wave radiation to be absorbed and warm the ice and water or be used in photosynthesis. On the other hand, the large oceanic heat flux from the warmer water below the surface in the deep Southern Ocean (p. 12) seems to retard growth of congelation ice so that the total thickness of ice tends to be less than in the Arctic. Antarctic ice consists mainly, 50–60%, of granular frazil ice as against the 80–95% of columnar congelation ice in the Arctic pack (Gow and Tucker in Smith 1990).

7.3 The microbiology of sea ice

Just as physical conditions differ, sea ice habitats in the Arctic and Antarctic are similar but not identical. Surface melt pools are common in the Arctic but infrequent in the Antarctic. The snow ice or infiltration ice layer assemblage (Fig. 7.2) has only been reported from the Antarctic. Communities within the ice are common in the Antarctic but less well developed in the Arctic. The increase in thickness of Antarctic ice by accumulation of frazil rising beneath it tends to incorporate plankton whereas the extension of stable Arctic pack downwards by growth of congelation ice does not. The microalgae reported from the sea ice of the two polar regions have generic resemblances but the species are not the same. However, Arctic and Antarctic can here be considered together by concentrating on general physiological aspects and ecological interrelations (Horner 1985; Melnikov 1997; Palmisano and Garrison in Friedmann 1993; Spindler in Hempel 1994; Vincent 1988).

7.3.1 The origins of sea ice communities

The assemblages found in sea ice are often termed *epontic* or *sea ice microbial communities* (SIMCO) and comprise microalgae, bacteria, protozoa, and small metazoans. Broadly, the microalgal assemblages, which are the most conspicuous, are classified into surface, interior, and bottom but the scheme in Fig. 7.2 gives a better picture. Bottom assemblages have been most studied, being accessible and attractive to SCUBA divers (for the technicalities of scientific diving under sea ice see Robinson *et al.* 1995). The surface of the ice and pools on it may be colonized from airborne propagules but the bulk of biota in the ice is likely to have a planktonic origin. The ice may begin with a considerable charge of organisms and organic matter since ice crystals collect and concentrate particles as they rise or as water is drawn through aggregations of frazil. Comparisons of the composition of interior communities of recently formed ice and that of the plankton beneath show close correspondence. Only a few species of diatoms (e.g. *Nitzschia* spp.), seem to be of benthic origin or have their distribution restricted to the ice edge. Of course, floras change with time as some species thrive and others perish in the altered environment. On the underside of the ice are mat/strand assemblages which hang down from congelation ice and are dominated by diatom species. This attached growth habit is not one which phytoplankton readily adopt and the species concerned are not abundant in the plankton. Another under-ice assemblage is found in the interstitial environment of loosely consolidated platelet ice. In this, the species are mainly common planktonic forms. Growths at the bottom of the ice are most characteristic of the less disturbed conditions afforded by land-fast ice.

7.3.2 Surface communities

Melt pools may be exposed or covered by a crust of ice. In the Arctic, where the pools are usually of freshwater, the flora, which includes species of snow algae (p. 58), is mostly of land origin (Melnikov 1997). In the Antarctic the algae seem mainly derived from the marine plankton, primarily small diatoms and flagellates but often the colonial *Phaeocystis* spp. (Palmisano and Garrison in Friedmann 1993). Some pools have almost no growth but in others the density may reach over 10^9 cells ml^{-1}. Standing crops of between 4 and 244 mg chlorophyll *a* m^{-2} are recorded. Microalgae from the surface of Weddell Sea pack ice achieve saturated rates of photosynthesis at higher irradiances, 100–150 μmol quanta $m^{-2}s^{-1}$, than do those within or at the bottom of the ice. The algae have attracted most attention but numerous heterotrophic protozoans and metazoa accompany them.

7.3.3 Communities within the ice

Diatoms are most prominent in these, often giving a distinct brown colour to the ice, but they are accompanied by *Phaeocystis* and other flagellates – both autotrophic and heterotrophic – protozoa, and bacteria.

Primary production in ice is of interest and it is necessary to consider the physiology of the photosynthetic organisms, particularly their capacity to utilize low irradiance levels and function at low temperatures and under osmotic stress (Lizotte and Sullivan 1992). Saturation irradiance for the sea ice diatoms *Amphiprora kufferathii* and *Pleurosigma* sp. has been determined as 25 and 13 μmol quanta $m^{-2} s^{-1}$, respectively. Shade adaptation (p. 46) of this order is not obligate for some Arctic ice algae. Temperature optima are in the range 2–7 °C and salinity optima around 30‰, somewhat below normal seawater. Osmotic acclimation is achieved by accumulation in the cells of substances such as the amino acid, proline, a mechanism similar to that used by algae in temperate waters. Cell division of some species can take place at 90‰ salinity and −5.5 °C. Nutrient concentrations in cores of sea ice vary between zero and two or three times that in the seawater beneath – during freezing, dissolved substances, such as nitrate, are concentrated in parallel with sodium chloride into the brine. Young ice usually has about the same concentrations of nutrients as the seawater whereas older ice may have either increased or depleted levels. Silicate, which is regenerated more slowly than nitrogen or phosphorus, may be expected to become exhausted first and this would account for an observed shift in the ratio of flagellates (which do not require silicate) to diatoms (which do) from 1 : 1 to 40 : 1 from July to September in Antarctic sea ice. The finding that the carbon fixed by algae growing in either Arctic or Antarctic ice goes largely into polysaccharides (60%), rather than into protein (10%), indicates nitrogen limitation. Confirmation that algae in brine channels do become limited by nitrogen supply is given by observations that algae growing in platelet ice, which is continually flushed with fresh seawater, have a normal cell protein content. Populations of ice algae supplied with additional inorganic nitrogen show a marked increase in standing stock.

Sea ice algae get little or no light during the winter but some species are able to survive in darkness for up to 84 days. Do they supplement photosynthesis by assimilation of organic substances? Such substances are certainly available from the bacterial degradation of particulate organic matter collected by frazil ice and of organisms which die in the brine channels. Furthermore, as with other substances, dissolved organic carbon is concentrated into the remaining liquid when freezing occurs. Calculations show that the supply of dissolved organic matter available in ice channels is of the same order as that of the organic matter produced by photosynthesis in sea ice and, while not all of it is necessarily available to microorganisms, it represents a not inconsiderable contribution. Some algae, including some sea ice species, are certainly capable of heterotrophic growth. Others have more limited heterotrophic ability. *Nitzschia sublineata* assimilates a labelled substrate, ^{14}C-serine, supplied at natural concentration in seawater, at a rate corresponding to less than 0.3% of its photosynthetic carbon fixation. This low rate of assimilation might maintain life during the winter months but, as for the similar problem with plankton algae (p. 188), the case is not proven. In competition with bacteria for organic substances, microalgae, because of their larger size and smaller surface/volume ratios, come a poor second.

Accompanying algae in sea ice are substantial populations of bacteria, phagotrophic flagellates including choanoflagellates, bodonids, euglenoids, and dinoflagellates, and ciliates. Sometimes, micrometazoa penetrate from the sub-ice assemblage but generally the microalgae are protected from grazing by larger predators. The basic energy sources for the heterotrophs must be organic detritus and dissolved organic matter contributed by the activities of photosynthetic algae and the heterotrophs themselves in addition to that derived from the original seawater. The bacteria, which reach rather higher densities than in seawater, are consumed by flagellates and ciliates and the whole community approximates to a closed, self-sustaining, system with regeneration by the heterotrophs of the mineral nutrients required by the autotrophs. The biomass can be considerable, sometimes reaching concentrations as high as 2 mg chlorophyll $a\,l^{-1}$. Primary production rates are variable in the range 0.1–4 mg C mg chlorophyll $a^{-1}\,h^{-1}$ with maximum rates per unit area of ice up to 35 mg $C\,m^{-2}\,day^{-1}$. Bacterial production rates are about 15% of this. There is strong evidence for algal/bacterial interdependence in this community (Becquevot *et al.* 1992; Palmisano and Garrison in Friedmann 1993). Numbers of algae and bacteria are positively correlated during spring growth and an investigation of the effects of temperature on uptake of carbon dioxide by algae and of thymidine by heterotrophic bacteria showed that responses were similar. The organization of the brine channel community, both at the biochemical and organism levels, is a challenging subject for investigation. *In situ* study with microprobes and endoscopes would be needed to avoid the disruption which accompanies ordinary sampling methods.

7.3.4 Communities on the undersurface of ice

The undersurface of sea ice, which is irregular and of varying texture (Fig. 7.4), provides niches for *sympagic* communities.

7.3.4.1 The Antarctic

The algal components here are mainly colonial or chain-forming diatoms. Although the irradiance which reaches them is less than 1% of that at the surface, the standing stock is often high, 520 mg $C\,m^{-2}$ having been found in McMurdo Sound under snow-covered ice and as much as 1076 mg $C\,m^{-2}$ in the same area under snow-free ice. The algae are strongly shade-adapted. Light saturation occurs around 20 μmol quanta $m^{-2}\,s^{-1}$ and photoinhibition

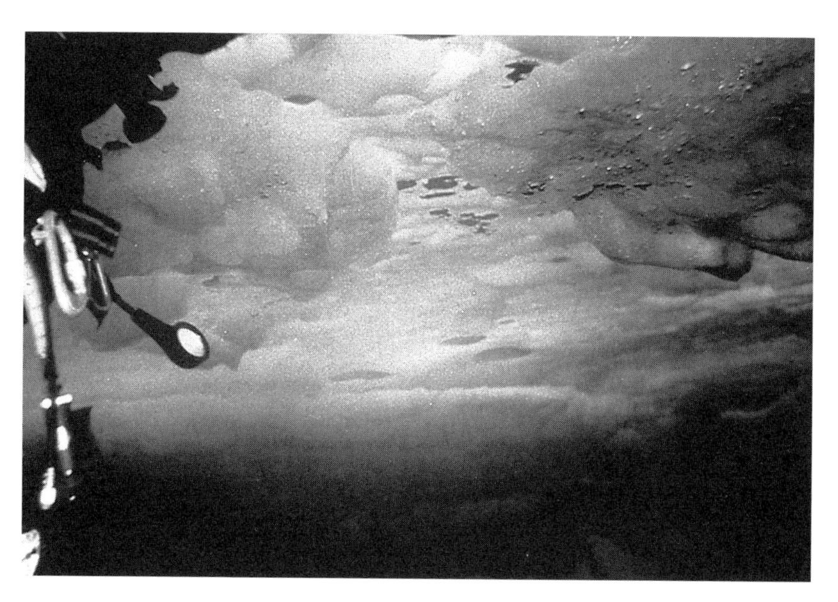

Fig. 7.4 Undersurface of pack ice, Weddell Sea. (Photo by courtesy of Dr C. Robinson.)

sometimes sets in at only 25 μmol quanta $m^{-2} s^{-1}$. The photosynthetic capacity is low, about 0.5 mg C mg chlorophyll $a^{-1} h^{-1}$, a value less than one fourth of that for temperate algae, and as much as 40% of the product escapes from the cells in soluble form. This is high compared with the 10% or so release which is usually found and is probably a response to stresses imposed by melting and dilution of samples. Under natural conditions, material released in this way would be quickly taken up by bacteria. The general relationships of the microorganisms in this community are presumably much the same as those in the interior of the ice but exchange of nutrients, organic matter, and organisms with the water column are more extensive. Both adults and larvae of copepods, together with mites and amphipods, are found in the sympagic community of the Antarctic, feeding off the algae and protozoa. There is also grazing and disturbance by larger pelagic animals.

A localized, but important, undersurface ice habitat is the platelet layer found under fast ice in Antarctica. Ice platelets are formed in supercooled water flowing from under ice shelves and float upwards to accumulate in layers up to 2 m thick. Algae and bacteria grow well at the interface between this layer and the fast ice above and, although nutrients may become limiting, there is protection from grazers and standing stocks of up to 190 mg chlorophyll $a\,m^{-2}$ have been recorded. The light reaching the platelet layer may be about 3% of that incident at the surface level but photosynthesis by the algae, mainly diatoms, is saturated at about 6–13 μmol quanta $m^{-2} s^{-1}$, indicating that they are acclimatized to the light environment (Robinson *et al.* 1995).

Associated with the sympagic community, preying on it but not part of it, is the *cryopelagic* community, composed of adult and larval stages of zooplankton and some fish. Some graze off the algae on the undersurface of the ice. These include amphipods, larval euphausiids, and copepods. The presence of *Euphausia superba* (krill) larvae is of particular interest. Krill is primarily planktonic and so more appropriately considered in the next chapter. However, it is mainly distributed in a zone of the Southern Ocean between the summer and winter limits of the pack and is frequently found below ice floes in the spring. Only recently have observations by SCUBA divers and remotely operated vehicles shown that its larvae are sometimes present in quantity on the undersurface of ice, feeding on the algae growing there. Whereas adult krill can tolerate starvation for over 200 days, older larvae can only survive for about 60 days without food. Overwintering beneath ice not only provides larvae with the food they need but also gives protection from predators. Although krill normally feeds by filtering out phytoplankton, it can adapt its feeding mechanism to capture algae scraped off the ice surface. More observations will be required to establish just how important overwintering under ice is in the annual energy budget of krill populations. Among secondary consumers in the cryopelagic community a dominant species is silver fish, *Pleurogramma antarcticum*, in its larval stage, which takes mainly nauplii and copepods. Another small pelagic fish, without a common name, is *Pagothenia borchgrevinki*. This is abundant as schools frequenting the few metres of water immediately under the ice and has a varied diet which includes *Pleurogramma* larvae, amphipods, copepods, and euphausiids. It is protected against freezing by antifreeze glycopeptides in its body fluids (p. 37).

7.3.4.2 The Arctic

As in the Antarctic, diatoms predominate in the sympagic communities but the species are largely different. As well as planktonic forms, there are characteristic species, such as *Nitzschia frigida*, which dominates sub-ice assemblages in the central Arctic Ocean and also occurs in the Antarctic, and *Melosira arctica*, which forms large mucilaginous masses (Hegseth 1992; Gutt 1995; Melnikov 1997). Nutrient concentrations in ice-covered regions in the Barents Sea are high in spring and analysis of protein/carbohydrate ratios indicates that the algae are not deficient in nitrogen or phosphorus. Diatoms *in situ* are shade-adapted with maximum rates of photosynthesis at about 50 μmol quanta $m^{-2} s^{-1}$ but in laboratory culture *N. frigida* adapts and maintains maximum rates up to 400 μmol quanta $m^{-2} s^{-1}$. Other species are not so adaptable. By late May there is light limitation as a result of self-shading. Growth rates, varying between 0.15 and 0.8 divisions per day, are in the same range as found for Antarctic species.

The Arctic has a more diverse sympagic fauna than the Antarctic, with harpacticoid and cyclopoid copepods, larvae of benthic invertebrates, and, often as the dominant forms, nematodes, which have not been reported from Antarctic ice (Melnikov 1997). The fauna of the Barents Sea ice is dominated by the amphipods *Apherusa glacialis*, *Onisimus* sp., and *Gammarus wilkitzkii*, which seem to be autochthonous sympagic forms dependent on permanent ice cover (Lønne and Gulliksen 1991). Less important forms include polychaetes, copepods, decapods, and the 'sea butterfly', *Limacina* sp. The amphipods together with the polar cod, *Boreogadus saida*, which frequents the underside of the ice, are important food sources for seabirds and marine mammals. The density of the fauna is variable, between zero and 2 g alcohol-preserved weights m^{-2} with values less than 0.001 $g\,m^{-2}$ preponderating. These densities seem typical of the seasonal ice zone but in older ice values increase and a total biomass of 36 $g\,m^{-2}$ has been reported from near Franz Joseph Land (*c.* 81°N 50°E). Seasonal ice has to be recolonized every year, resulting in lower faunal densities than in perennial ice. It seems that the movement of ice from the Polar Basin through the Fram Strait leads to an annual loss of 7×10^5 tonnes of biomass from the perennial ice zone. Seasonal ice off the northeast Greenland coast carries heavy but patchy growths of *Melosira arctica* with no significant associated fauna (Gutt 1995). After ice melt this diatom sinks to the bottom where it may provide a substantial input into the benthic system. Although perennial Arctic sea ice is in a state of continual vertical flux the organisms inhabiting it maintain themselves as a distinct ecosystem (Melnikov 1997).

7.4 The epifauna of the ice

The birds and mammals which to the casual seafarer seem most characteristic of sea ice are scarcely dependent for food on the ice communities just described but have evolved foraging techniques adapted to the physical nature of the ice and may make use of it as a comparatively safe breeding ground. Arctic and Antarctic are here best dealt with separately, the species involved being different.

7.4.1 The Antarctic

The emperor penguin (Williams 1995; Ancel *et al.* 1997) is superbly adapted to existence on sea ice and has had perhaps 40 million years to evolve to this state. The total population lies between 135 000 and 200 000, distributed in about 30 breeding colonies around the continent. It feeds by pursuit diving for fish, squid, and krill, going to depths of as much as 265 m. Living entirely on the ice when not at sea, it gets round the difficulty of nesting by incubating its single egg balanced on top of its feet and enveloped in a feathered fold of abdominal skin. This needs a flat surface and breeding sites are on level sea ice in sheltered situations (Fig. 7.5). Eggs are laid in the summer and the females then go off to sea to recuperate. Incubation is carried out over winter by the males, so that the chicks are fledged by midsummer in time to take to sea and allow that adults sufficient pause to

Fig. 7.5 Emperor penguins, Brunt Ice Shelf, January. (Photo G. E. Fogg.)

moult and return to breeding condition. The young birds travel north on floes as the ice breaks up.

The emperor has adaptations for survival through the Antarctic night when temperatures on the ice may fall to $-48\,°C$ and winds reach $180\,km\,h^{-1}$. First, the bird is large, weighing up to 46 kg, and thus has a small surface/volume ratio which is further minimized by having flippers and bill about 25% smaller in proportion than other penguins. Insulation is provided by double-layered high density feathers and a 2.3-cm thick layer of subcutaneous fat. The thermal conductivity of fat is a quarter of that of water but the feathers, with the air layer which they entrap, provide more than 80% of the thermal insulation, even in a breeze. Flippers and feet have vascular counter-current heat exchange systems which reduce heat loss by using the heat of the outgoing blood to warm up the returning flow instead of being lost to the air. These features allow the metabolic rate to remain at normal level down to a critical temperature of $-10\,°C$, below which it must be increased if body temperature is to be maintained. A remarkable behavioural characteristic enables heat and energy reserves to be conserved below this temperature. Unlike other penguins, which in common with other birds have strong territorial instincts, the male emperor does not object to being in close proximity to his fellows and under adverse conditions joins a huddle which may contain as many as 5000 birds, packed 10 to the m^2. The whole huddle moves slowly downwind as birds on the windward side move along the flanks and re-enter it on the lee side. If the emperors are startled and raise their heads, steam can be seen rising from the huddle. It is estimated that by this means the daily loss of body weight is cut by 25–50%. At the end of the winter the male emperor, which by then has lost up to 40% of his summer body weight, still retains enough reserves to produce a nutritious secretion which keeps its chick going until the female arrives from the sea with food.

At present, emperor populations are stable or expanding slowly but it seems likely that they could be particularly sensitive to changes in sea ice patterns. For example, early melting and break-up could lead to the loss of an entire year's brood because colonies would be disrupted, adults could no longer find their young, and exposure to sea spray would be lethal for chicks with down feathers.

The Adélie penguin, *Pygoscelis adeliae* (Williams 1995), the most numerous of Antarctic penguins and frequently seen in the pack ice, nests on land in the early spring or summer and avoids the extreme conditions of winter on sea ice by going north to open pack ice 150–650 km from the continent. Birds tracked by satellite telemetry carried out winter migrations of as much as 5000 km. On their return they may trudge several hundred km over ice to their breeding grounds, which are positioned where open water will eventually arrive at the time when chicks need to be fed. Less than two months are available between hatching and the time when the sea begins to freeze again, forcing fledglings to take to the water. Adélies feed on krill and other crustaceans. Observations on birds fitted with miniature depth gauges show that occasionally they dive as deep as 170 m but usually stay above 45 m. Unlike the emperor, the Adélie has a range extending north to regions where it need have little involvement with sea ice. Both these birds are preyed on by leopard seals (*Hydrurga leptonyx*), which also frequent the pack, but they are largely immune from attack when on the ice although the seals sometimes penetrate thin ice from below to take Adélies walking on top. Neither emperor nor Adélie could exist were there carnivores similar to polar bears or Arctic foxes in the Antarctic. Flighted birds do not breed on sea ice and visit it only to forage.

There are four species of ice-breeding seals in the Antarctic – Weddell (*Leptonychotes weddelli*), crabeater (*Lobodon carcinophagus*), Ross (*Ommatophoca rossii*), and, just mentioned, the leopard (King 1983; Laws in Laws 1984; Knox 1994; Fig. 8.5). The crabeater is the most abundant seal in the world, with a total population of the order of 30 million. The Ross seal perhaps numbers 200 000 but is rarely seen. The Weddell and crabeater seals afford a contrast in use of the ice. The Weddell breeds on the nearshore ice, selecting areas where cracks are perennial features. In summer, outside the breeding season, they move to the outer limit of the fast ice and the inner zones of the pack. Here they are usually seen singly and their mean density has been estimated as 0.14 km^{-2}. The Weddell seal extends as far north as there is reliable pack ice in the winter. Its food is mostly fish and cephalopods with some invertebrates but only a small amount of krill.

Diving, which has been intensively investigated, using attached time-depth recorders in its nearshore habitat, is of two types. Deep 'working' dives of short duration, 5–25 minutes, go to depths of 200–400 m, whereas long, 20–73-minute, 'exploratory' dives go to depths which do not take the seal out of sight of the surface. During the winter the Weddell seal remains under the ice, rarely hauling out (it must be remembered that even at $-1.9\,°C$ the water is warm compared with the air). Breathing holes are kept open by rasping away the ice, using the well-developed incisor and canine teeth. The eventual wearing away of these teeth seems to be a common cause of death. In spring, pupping colonies assemble around breathing holes and most pups are born, on the ice, by mid-November. Pups are suckled for a month or two but enter the water in their second week under their mothers' care. Mating takes place underwater around midsummer. The Weddell seal is polygynous but a male is only able to exert his authority over the females which frequent one particular breathing hole. These are usually not more than ten.

Crabeater seals are similar in size to Weddell seals but occupy a different niche in the sea ice habitat, preferring pack to fast ice (Fig. 8.5). They are more gregarious than Weddell seals and are sometimes found in summer in aggregations of as many as 600 within a radius of 5 km. The difficulty of counting animals which are spread out over an enormous area of sea ice scarcely needs pointing out. Census has been carried out by combined use of aerial photography, ice-breakers, and helicopters to sample strips orientated north–south to penetrate the ice to a maximum distance. The data obtained indicate mean densities of about $4.8\,km^{-2}$ in summer and $0.5\,km^{-2}$ in winter and spring. Their usual food is krill rather than larger animals such as crabs – which, anyway, are not found in Antarctic pack ice. Their molar teeth, which have prominent cusps, are well adapted for filtering out these shrimp-sized organisms and they seem to dive only to the relatively shallow depths where krill are most abundant (Fig. 8.7). Pupping takes place on the ice from late September until early November but, because of their inaccessibility in the pack at this time, information about breeding behaviour is sparse. Unlike the Weddell seal and those seals which breed on land, the crabeater is monogynous (p. 172).

The difference between the nearshore fast ice habitat and the offshore drifting ice leads to contrasts in the biology of the two species. The crabeater, living in an environment which is constantly changing and therefore clean, rarely has parasites whereas the Weddell is usually infected with a variety of pests. Using electrophoretic techniques to characterize proteins, it is found that the circumpolar population of the crabeater is homogeneous whereas the Weddell has at least two isolated breeding populations, presumably because it is less widely roving.

A major predator on both of these seals is the killer whale or orca (*Orcinus orca*). This has world-wide distribution but concentrates in the colder waters of both hemispheres. It moves about in packs, sometimes of 30 or 40 animals, which are frequently seen in pools amid the pack ice. Hunting is a concerted activity and they are reputed to tip seals off ice-floes into the water. Man has never been a major predator on the seals of the Antarctic sea ice.

7.4.2 The Arctic

Since the flightless great auk (*Alca impennis*), which nested on offshore skerries, became extinct on 4 June 1844 there has been nothing penguin-like in the Arctic. Presumably, the great auk foraged at the ice edge as far away as possible from polar bears. As in the Antarctic, flighted birds, such as the existing auks, use the pack ice as feeding ground but do not breed on it (p. 204). Polynyas seem particularly important for the wintering of birds in high latitudes. Most of the population of Ross's gulls (*Rhodostethia rosea*) seem to winter around them, for example, at the Velikaya Sibirskaya polynya in the Russian Arctic. Regular wintering populations of Brünnich's guillemot (*Uria lomvia*), little auks (*Alle alle*), and long-tailed ducks (*Clangula hyemalis*) are found in the vicinity of polynyas north of the edge of the pack ice. Polynyas determine migration routes and the positioning of nesting grounds on neighbouring coasts (Fig. 7.9; Sage 1986).

Six species of Arctic seal use sea ice as a habitat (King 1983; Sage 1986; Pielou 1994). The walrus (p. 143) feeds on benthos and therefore prefers shallow waters but, although it may congregate in large numbers on beaches, it characteristically lives in the pack and at ice edges, breaking through 20 cm of ice by butting with its head. It mates around midwinter among the ice, copulating under water with one bull serving perhaps 15 females. The spotted (*Phoca largha*), ribbon (*Phoca fasciata*), hooded (*Cystophora cristata*), and bearded (*Eringnathus barbatus*) seals all frequent drifting ice. The bearded seal prefers shallow waters free of fast ice, with moving floes and open leads, but it can keep open breathing holes by means of the strong claws on its fore flippers. The ringed seal (*Phoca hispida*), the commonest seal in the Arctic, with a total population of between 6 and 7 million, is found in open water in the fast ice but rarely on floating pack ice or in the open sea. In winter, adults stay under the ice in bays and fjords. The younger ones mostly stay further out at the edge of the fast ice. In summer most lie out on the ice basking in the sun, moulting and fasting. Like the bearded seal the ringed seal keeps open breathing holes, which may extend through ice 2 m thick, by abrasion with its flipper claws. A low dome of ice, about 4 cm high, marks the hole at the surface and the snow which accumulates round this, if it is sufficiently deep, may be hollowed out and used as a lair. Larger lairs are constructed by pregnant females and the pups are born in them in spring. These lairs, with their adjacent breathing holes giving access to the sea, not only give some protection against polar bears and Arctic foxes but provide shelter from cold and wind. The pup, which has not got the insulating layer of blubber possessed by adults, must derive appreciable warmth from its mother in the confines of the lair.

The walrus is polygynous and so, probably, is the ribbon seal. The male ringed seal mates with three or fewer females. The spotted, hooded, and bearded seals are monogynous. There is thus the same tendency towards monogyny in ice-breeding seals as is seen in the Antarctic. This is accompanied by near equality in size of the two sexes. The reasons for this are not altogether clear. The selective pressures operating against the formation of female–pup aggregations defended by a dominant male which are characteristic of land-breeding seals may arise from the difficulties of defending a group in the unstable environment of the pack, from aquatic mating, and, in the Arctic, from the presence of large predators on the ice.

Man has headed the list of predators but for several thousand years was in balance with the sea ice ecosystem rather than a major disturbing influence. Hunting in the sea ice for seals developed particularly with palaeo-Eskimos who moved into the northwest Canadian Arctic and western Greenland 2000 years BP and became the present-day Inuit (p. 214). Sophisticated means of harvesting food from inshore waters were developed. In winter, hunters moved over the inshore ice by dog-sledge and, where the ice was clear, breathing holes were approached with feet muffled in polar bear skin shoes and watched until a seal appeared and could be harpooned (Fig. 7.6).

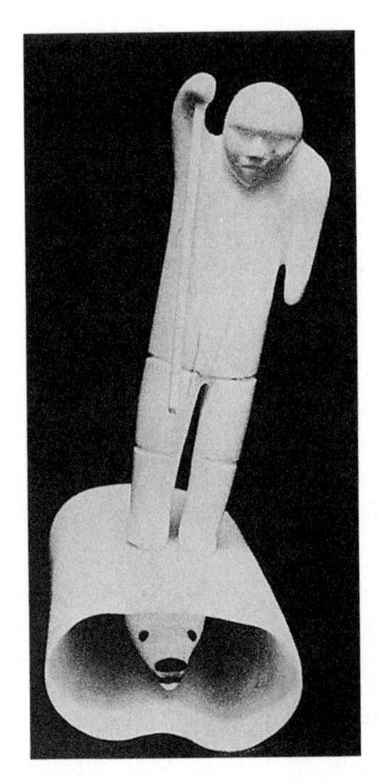

Fig. 7.6 Carving in walrus ivory of Inuit seal hunter by Fabien Oogark, *c.* 1959. (From G. Swinton 1972 *Sculpture of the Eskimo*, M. F. Feheley Arts Co.)

When snow obscured the holes, dogs were used to sniff them out. Winter sealing was combined with fishing through holes in the ice. In spring, when the seals bask on the ice, a hunter might creep up, lying down to imitate a seal if sighted by his prey. In summer, the sealskin kyak provided a swift and silent means of approach through the pack. Walruses and, especially, ringed seals were taken around Greenland and the Northwestern Territories. Now canoes with outboard motors are used and the seals shot with high-powered rifles. Many thousands of ringed seals are caught each year but the walrus is now protected, apart from small numbers which local populations are allowed to take. The bearded seal is of great importance to coastal natives in Alaska and around the Bering, Chukchi, and Beaufort Seas. In 18 months in 1977–8 rather more than 8000 were taken around Alaska. In the Bering and Chukchi Sea areas they are caught not only by the natives but by Russian commercial sealers (Sugden 1982; McGhee in Ives and Sugden 1995).

Although they often produce their cubs on land and are closely related to the terrestrial grizzly bear (*Ursus arctos*), polar bears (*U. maritimus*; Sage 1986; Pielou 1994) are essentially sea ice animals. Apart from occasionally eating berries or grasses, and a recently developed habit of raiding rubbish dumps, they are dependent on the sea for food (Fig. 7.7). They are well adapted to survive low temperatures by virtue of their large size and fur which has six or seven times the insulating capacity of the clothing normally worn by humans at rest in an ambient temperature of 20 °C. The fur is colourless and transparent to short-wave radiation, which is absorbed by their black skin, but opaque to infrared – hence polar bears are not detected by infrared imaging equipment. The total population of around 25 000 has a

Fig. 7.7 Swimming polar bear. (Photo by courtesy of F. Bruemmer.)

circumpolar distribution but sub-populations tend to keep to particular localities. However, tracking by means of satellite-monitored radio collars and aerial surveys have shown that some individuals undertake long journeys, such as from Svalbard to the Russian Arctic. Individuals are sparsely scattered as far north as 88° and as much as 200–300 km from land. Two kinds of ice seem to be preferred; one, ice with nearly complete cover but with sufficient movement to create leads, and the other, the ice edge where leads are wide. Apart from this they are found on flat land-fast ice with stable pressure ridges, drifted with snow, suitable for establishing lairs. Polar bears, which have partially webbed feet and are strong swimmers (Fig. 7.7), feed on seals, fish, seabirds, kelp, and, occasionally, stranded whales. They catch seals by lying quietly in wait by their breathing holes and also attack female ringed seals and pups in their lairs. These bears have in turn been hunted by man but are now protected. The survival of cubs shows great variation, depending on the level of fat storage in the females prior to giving birth, and on the availability of ringed seal pups afterwards. This in turn is presumably related to fluctuations in oceanographic or climatic factors (Stirling and Lunn in Woodin and Marquiss 1997).

The Arctic fox (*Alopex lagopus*; Sage 1986; Pielou 1994) although chiefly an animal of the land, is a scavenger of the leftovers of polar bears and is itself an important predator on pups of the ringed seal, catching them, like the polar bear, in their birth lairs. From a three-year study in the Canadian Arctic an average pup predation level of 26% by Arctic foxes on nearshore ice was estimated.

7.5 The ice edge

The ice edge is not simply a line where sea ice and its associated organisms disappear to be replaced by open water and pelagic biota. It is a dynamic zone, constantly shifting in position and interacting with the atmosphere to produce characteristic weather (Vincent 1988; Ainley and DeMaster in Smith 1990). The physical processes taking place modify biological processes to a profound extent. Because the situation in the Antarctic is somewhat more straightforward, this will be discussed first.

A large amount of melt water is produced in a relatively short period in the spring and summer. Of the $20 \times 10^6\,km^2$ of ice which cover the Southern Ocean by the end of winter about $15 \times 10^6\,km^2$ melt during the summer, implying that a layer of low salinity water of about the same thickness as the ice is produced over the same area. Under turbulent conditions this may be mixed into the water column but the less dense freshwater has a powerful stabilizing effect working against this. Because of the short fetch of wind over the water between floes, down-mixing is minimized. Lenses of nearly freshwater, with salinities as low as 0.03‰, occur near melting ice edges. The marginal zone in which melting takes place may be several hundred km wide but, of course it is not fixed and travels polewards each summer. The density gradients at the fronts between saline and less saline waters set up both vertical circulation and horizontal currents paralleling the fronts (p. 24).

Organisms released from melting ice undergo osmotic shock as the brine in which they were living is diluted and there is also an abrupt increase in irradiance to cope with. Some forms may be killed and those which survive may suffer temporary alterations in cell membrane permeability, causing them to release soluble cell components, making more organic material available to heterotrophs. During the winter there was deep vertical mixing, associated with cooling and ice formation, to perhaps 125 m below the ice. This, together with low light penetration, kept phytoplankton production at a low level. The stabilization accompanying melting confines mixing to the surface 10 m or so and conditions become much more favourable for photosynthesis. The stabilized area may extend 100 km into the open sea but is not necessarily uniformly stratified. There may be a mosaic of mixed and stratified patches. Local hydrographic features, such as fronts, may override the ice edge effect. The inoculum for plankton growth may come from the ice, opportunistic species growing to dominate the community until mixing brings in species able to compete with them. For example, in the marginal ice community off southern Victoria Land the predominant phytoplankton species is *Nitzschia curta*, a member of the local sea ice community. It is active in photosynthesis under these conditions. On the other hand, there is a tendency for algae released from ice to aggregate. Aggregates sediment more rapidly than individual cells and are more readily grazed and so are eliminated to a greater extent. This happens in the northwestern Weddell Sea, where their contribution to the phytoplankton is negligible. *Phaeocystis* is often abundant at the ice edge and responsible for chlorophyll concentrations as high as 14 mg m^{-3}. In this area the bulk, around 74%, of the photosynthetic biomass is in nanophytoplankton like *Phaeocystis*, less than 20 μm in size. In the Southern Ocean nutrient concentrations are nearly always ample to support dense growths. Often, the density of the phytoplankton is such as to show up in satellite images as a belt along the ice edge. Rates of primary production are correspondingly high with a mean value as much as 1.76 g C m^{-2} day^{-1} in spring, but declining to about a quarter of this by autumn and back to less than a tenth in winter. This compares with 0.36 and 0.87 g C m^{-2} day^{-1}, mean spring values for water under close pack and in open ocean, respectively (Mathot *et al.* 1992).

Roughly coextensive with the zone of phytoplankton concentration is one with high bacterial densities (Fig. 7.8). This growth is presumably sustained by organic matter released first by melting of the ice and then by algae and zooplankton. In the Weddell Sea, bacterial biomass is about 16% of that of the phytoplankton and 7% of the total microbial biomass. Protozoa, such as choanoflagellates, phagotrophic dinoflagellates, ciliates, and amoebae, contribute rather more to total biomass, 23%, than do bacteria. Nanoplankton, bacteria, and protozoa are the constituents of the ultraplanktonic

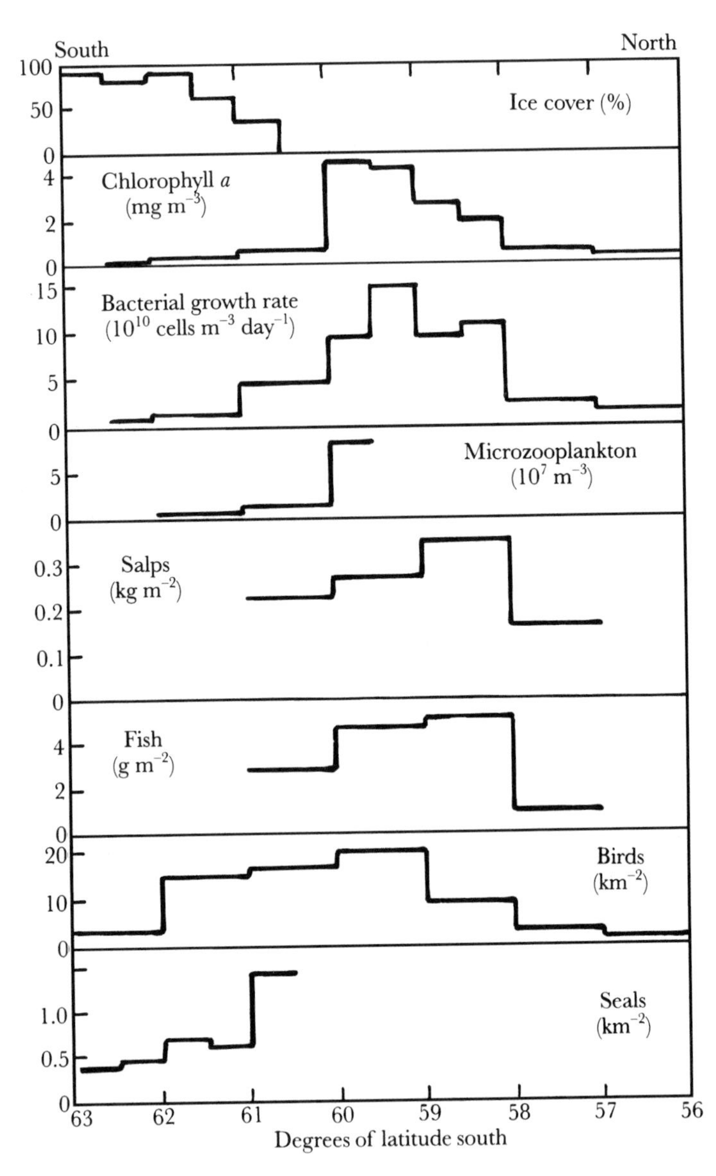

Fig. 7.8 Distribution of organisms along a south–north transect across the marginal ice zone in the Weddell Sea, November 1983. (Redrawn from Vincent 1988, courtesy of the author and Cambridge University Press.)

community to be described in the next chapter. There are close interrelations between the population dynamics of these organisms with the phagotrophs ingesting perhaps half of the daily primary production with concomitant recycling of mineral nutrients.

There has been debate about how much energy the ultraplankton contributes to higher trophic levels. Whether the consensus that it is not much applies to the marginal ice zone is uncertain. There has been a suggestion that comparatively large zooplankton forms, such as krill, can benefit from food concentration by protozoa such as tintinnids and choanoflagellates. Be this as it may, higher trophic levels have greater biomass and activity in the marginal ice zone paralleling those in the lower levels (Fig. 7.8). There is greater productivity at the microplankton level, sedimentation of organic detritus fuels more benthic production, and decaying ice floes provide refuges for the larger crustacea which attract fish, seabirds, and seals.

As well as by changes in biomass, the marginal zone is marked by differences in species, particularly of seabirds. The pack ice community is dominated in terms of biomass by emperor, chinstrap (*Pygoscelis antarctica*), and Adélie penguins, and in numbers by snow (*Pagodroma nivea*), and Antarctic petrels (*Thalassoica antarctica*) (Fig. 8.5). The highest density of this assemblage lies around 7–10 km north of the ice edge, which it follows as it moves south to take advantage of the surge in production. Further north of this the abundant seabirds are southern fulmars (*Fulmarus glacialoides*), and cape pigeons (*Daption capense*). There is a difference in the prey of these two groups of birds. Within the pack ice, crustacea of the genera *Pasiphaea* and *Eurythenes* are taken. North of the ice edge, the prey is mainly krill and small lantern fish (*Electrona antarctica* and *Gymnoscopelus braueri*), which rise to the surface at night to feed on krill.

The processes at ice edges in the Arctic are not as clear-cut. In the Southern Ocean the marginal ice zone lies over deep water, whereas in the Arctic there is year-round ice cover in the deep water regions and the ice margins lie over broad, shallow, continental shelves. The melt season is shorter than in the Antarctic and ice margins more subject to disturbance by currents. In the Chukchi Sea, for example, the summer ice edge system has upper and lower layer fronts, the upper arising from melt water and the lower, in a depth of only 50 m, marking the boundary between cold Chukchi Sea water and northern flowing warmer water intruding from the Bering Sea. The extent of the marginal ice zone depends on mesoscale dynamics and it is usually about 50 km wide as compared with 250 km in the Antarctic, where large-scale dynamics have more effect.

The general paucity of microbial growth in the interior of Arctic ice means that on melting there is little release of inoculum into the sea. Whereas phytoplankton growth at the Antarctic ice edge is sustained by high nutrient concentrations, the lower concentrations in Arctic seawater are quickly depleted and wind-driven upwelling of nutrient-rich water is necessary to support significant algal growth. Phytoplankton growth in the productive shallow-water shelf areas often masks increased production at the ice edge.

The marginal ice zones of the Arctic are, nevertheless, important for regional biological activity. Many species of seabirds and mammals use the ice edges as migration routes in the spring, depending on the reliable food supply they offer (Fig. 7.9). Availability of prey is reflected, too, in the

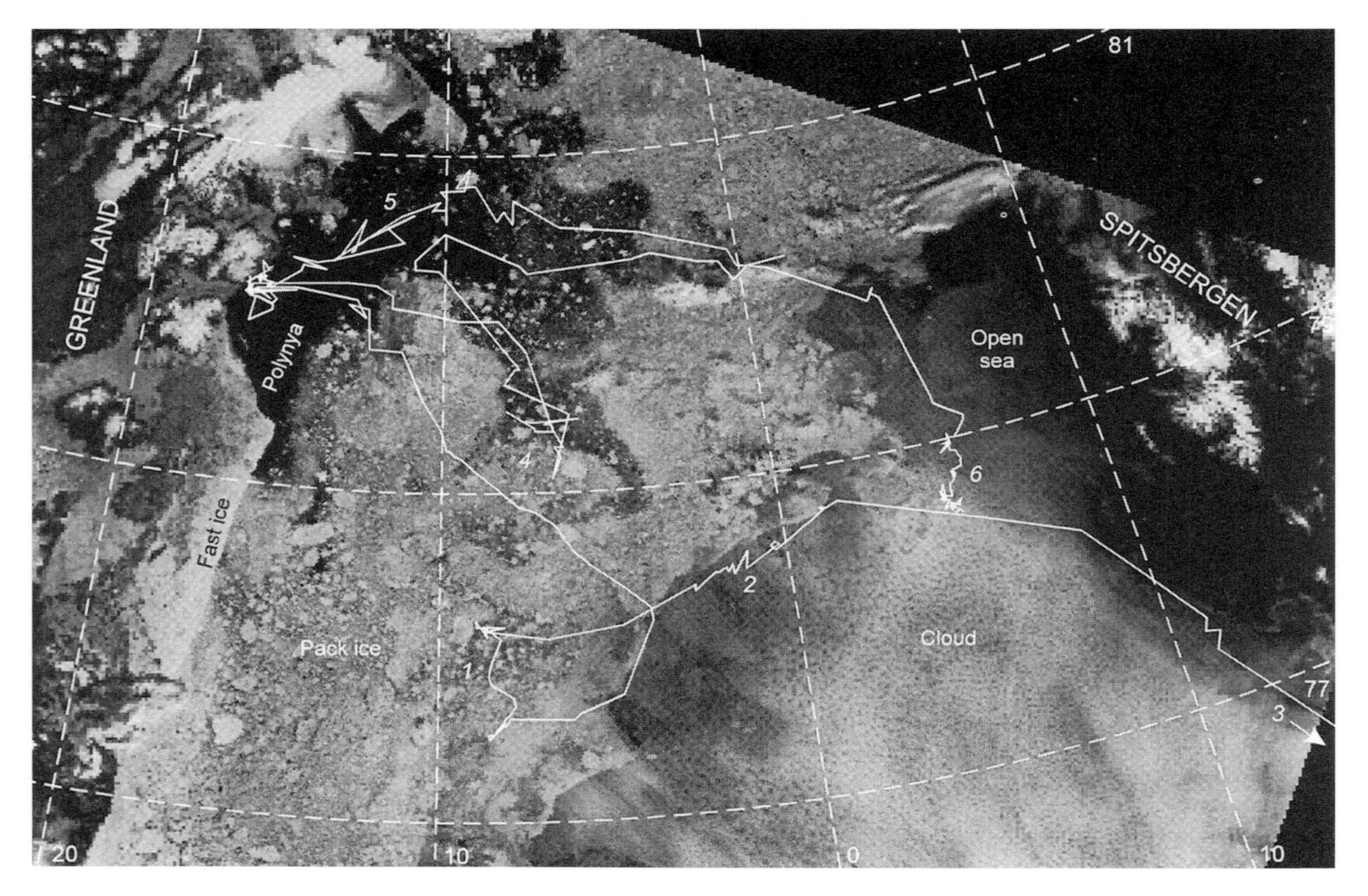

Fig. 7.9 Movements of fulmars in relation to ice patterns in the Greenland Sea (NOAA satellite image from 16 July 1993). The numbers mark resting/foraging areas. (From Falk and Møller 1995, courtesy of the authors and Springer-Verlag, Berlin.)

density of seabirds, which in the marginal ice zone of the Bering Sea is estimated as 500 individuals km^{-2} as compared with 0.1 km^{-2} in the ice to the north and 10 km^{-2} in adjacent open water. There are also effects on species distribution, as in the Antarctic. Appreciable numbers of seabirds are regularly found in the summer pack ice, feeding on Arctic cod and crustaceans, but only Ross's gull, ivory gull (*Pagophila eburnea*), and the black guillemot (*Cepphus grylle*) are characteristic of the pack and dependent on ice-associated fauna for the bulk of their food. The two gulls are rarely seen over the open sea although some races of the black guillemot have distributions going into temperate regions. Just south of the ice edge, however, a variety of species is to be found. In the Barents Sea, for example, glaucous gulls (*Larus hyperboreus*), herring gulls (*L. argentatus*), Brünnich's guillemot, and little auks, move in from open water to congregate at the ice edge. Brünnich's guillemot, one of the most numerous seabirds in the northern hemisphere, produces young which abandon the nesting ledges before they are able to fly. The ice zone around the coast is thus a barrier to them and breeding colonies are perforce located to the south of the summer ice margin. In the Davis Strait and the Labrador Sea ivory gulls are absent only during the short ice-free period in late summer. During March the largest numbers are at the ice edge near the breeding areas of the hooded seal, where they feed on the afterbirth. In April and May they prey particularly on lantern fishes near the ice edge.

7.6 The wider significance of sea ice

Sea ice supports a variety of communities, some of which can be surprisingly productive. Given the vast extent of sea ice it seems possible that its contribution to world primary production may be appreciable (Legendre *et al.* 1992). A question of importance in the context of global warming, is how much of the carbon fixed by sea ice communities is withdrawn from circulation and thus helps to counter the 'greenhouse effect'. The heterogeneity of sea ice, its great seasonal and annual variability, and the logistic and technical problems of studying it, make estimates difficult. Table 7.1 summarizes some made in 1992, on the basis of the few available data, by an international working group. It will be noticed at once that direct comparisons between Arctic and Antarctic are not entirely possible. Scientists working in the two regions have not always used comparable methods and have often had different objectives in mind. The various ecosystems are not equivalent; there is little real production in the interior of Arctic sea ice whereas in the Antarctic this is appreciable. The marginal ice zone is distinct in the Antarctic but in the Arctic it is best regarded as part of the shelf ecosystem. Nevertheless, the values given in the table have some consistency. Those for total production are similar for the two polar regions and the rates and standing crops in the different areas have the relative values that one would expect. Production associated with the ice itself is between 3.2 and 34% of the total in Arctic waters. In the Antarctic it is

Table 7.1 Microalgal and bacterial production in areas of the Southern Ocean south of *c.* 60°S and of the Arctic seas north of *c.* 65°N

	Uptake rate or stock	Area	Annual production
Antarctic			
Water column production at ice edge			
Phytoplankton	32–521 (R)	1.5–2.0	1.41
Bacteria	29–70 (R)	1.5–2.0	0.39
Water column production under ice cover			
Phytoplankton	168 (S)	5–10	0.034
Bacteria	8–20 (S)	5–10	0.403
Production associated with sea ice	80–32 000 (S)	0.265–20	0.63–0.70
Total			2.87–2.94
Arctic			
Water column production in shelf regions	27 000 (R)	4.9	1.32
Water column production in offshore regions	9000 (R)	8.2	0.73
Production associated with sea ice	600–10 000 (S)	0.6–7.0	0.09–0.73
Total			2.14–2.78

Carbon uptake (R) in $mg\,C\,m^{-2}\,day^{-1}$ (Antarctic), $year^{-1}$ (Arctic), standing stock (S) in $mg\,C\,m^{-2}$, areas in $10^6\,km^2$, production in 10^8 tonnes C $year^{-1}$.
Data from Legendre *et al.* (1992).

perhaps 2% of the total and the marginal ice zone contributes 4% of the total net primary production in the sea south of the Polar Front. The annual total net primary productivity of the world ocean – itself a subject of much uncertainty – is about 260×10^8 tons C. Primary production in sea ice is from 0.5–3% of this, a small, but not negligible proportion. These inputs sustain secondary production not only in the ice but in pelagic and benthic systems. However, primary production often happens in bursts and the organic matter of algal blooms may be in excess of herbivore capacity and consequently sediments in the forms of intact cells and faecal pellets. The accumulated organic matter released from Arctic multi-year ice when it melts may be mineralized quickly then, depending on circulation patterns, its carbon returns to the air or is retained at depth. Another possibility is that it may be converted to dissolved organic matter. As much as half of the organic matter produced in a bloom may be in the form of soluble substances which break down only slowly, perhaps having a half-life of a 100 years, but it is not known to what extent this happens in polar waters. It is surmised that in the Antarctic, areas in the Weddell and Ross Seas are sites of active 'biological pumping' of carbon into deep water, thus reducing atmospheric concentrations. The sedimentation of large diatoms from the marginal ice zone may provide one mechanism for the 'pump' but qnantification is still lacking.

The advance and retreat of sea ice imposes a corresponding pattern on the entire polar marine ecosystem through its effects on water temperature, penetration of light, and stability of the water column. These obviously affect levels of biological productivity, cycling of materials, and energy, and, as with krill (p. 197), life cycles and behaviour. Because most oceanographic expeditions have been confined to the summer months we know little about these effects of ice and the picture is complicated by the great variation from year to year in the extent, distribution, and duration of sea ice. Records of sea ice around the South Orkney Islands going back to 1903 have allowed analysis of this interannual variability (Murphy *et al.* 1995). Over this time there was a long-term decline in duration, with cycles having a period of less than 10 years being especially evident between the mid 1960s and 1990. Satellite data show that these local variations reflect larger-scale perturbations in the Weddell Sea, these in turn being part of a pattern which precesses clockwise around the Antarctic continent with a period of 7–9 years. Physical processes in both atmosphere and open ocean seem to be linked to this precession and one would expect that short- and long-term ecological changes are too. Local variations in primary production and effects at higher trophic levels, such as recruitment of marine invertebrates, in fact do show correlations with the dynamics of sea ice. On the global scale air–sea interchange of biogenic gases, such as carbon dioxide (p. 212) and dimethyl sulphide (p. 188), which have effects on climate, must be strongly influenced by variations in sea ice. All this points to a difficult but attractive field of research calling for collaboration of meteorologists, atmospheric chemists, oceanographers, and ecologists, both terrestrial and marine.

8 The open seas

8.1 Introduction

The ice-free areas of the polar seas do not differ in any fundamental way from the temperate seas with which they are confluent. They are colder and more tempestuous but these are only matters of degree and, although there are special features of hydrography, so there are in every other sea. Nevertheless, there are good reasons for including them as polar habitats. They interact with ice and the polar atmosphere, play important roles in polar thermal regimes (p. 8), and provide the organic production which sustains the animal life of both Arctic and Antarctic coastal regions. In detail they have their own biological characteristics. Study of polar marine biology has lagged behind that in other sea areas but it is evident that the Arctic seas and the Southern Ocean play crucial parts in global processes and that knowledge of their biological, as well as physical, oceanography is essential if we are to understand and manage our environment. The hydrographical and physical background has already been outlined (p. 8) and this chapter is concerned with the ecology.

8.2 The plankton

Plankton includes all those organisms, from viruses, bacteria, and microalgae to small animals, which live freely in the water column and whose movements are determined more by currents and eddies than by their own motive powers.

8.2.1 The ultraplankton

Plankton is of two distinct types, the ultraplankton and the microplankton, differing in size, form, physiological functions, and ecological roles (Fogg 1991*a*). The *ultraplankton* is a community of microorganisms with linear dimensions of less than 20 μm, including *nanoplankton*, from 2 to 20 μm, and *picoplankton*, 0.2 to 2 μm. These dimensions are small in comparison with the smallest eddies and the only motion to which the cells are directly exposed is that of molecular diffusion (p. 26). The ultraplankton occupies what has been called the *microlitersphere* and has been described functionally as the *microbial loop*.

Most information about the photosynthetic microorganisms of the polar seas has been obtained by sampling with the traditional plankton net of fine silk with a mesh size of about 40 μm. This retains perhaps only 10% of the total phytoplankton. To obtain correct representative samples including picoplankton it is necessary to collect water and centrifuge it or filter cells from it using a micropore membrane. Determinations of primary productivity involving filtration are sometimes ambiguous because the pore size of the filters used are not specified so that one is not sure whether picoplankton is included or not. In the northern Foxe Basin (*c.* 68°N 80°W) between 10 and 70% of the photosynthetically active flora passed through a 1 μm filter (i.e. was largely picoplanktonic). In Antarctic waters nanoplankton (photosynthetic flagellates) contribute about 70% of the total phytoplankton biomass in the Ross Sea. Such observations show that polar seas resemble other oceanic areas, in which ultraplankton is ubiquitous and the major primary producer under oligotrophic conditions.

The picophytoplankton includes prokaryotic forms – cyanobacteria and prochlorophytes – and unicellular eukaryotes belonging to the Chlorophyceae and Prasinophyceae. The nanophytoplankton comprises principally flagellates, belonging to various groups, with some diatoms. Viruses, bacteria, phagotrophic flagellates, and ciliates together provide heterotrophic activity which balances the phototrophic activity of the ultraphytoplankton. The bacterioplankton of polar seas is of about the same density as that of temperate waters, usually within an order of magnitude of 10^6 cells ml^{-1} and correlated with phytoplankton biomass. It consists of both psychrophiles with optimal growth below 10 °C and maintaining active metabolism at temperatures around zero, and psychrotolerant bacteria showing greater low temperature inhibition. Organic substrate (e.g. glucose) uptake in samples from McMurdo Sound (77°30′S) was increased by about 30% when the temperature was raised from ambient, −1.8°, to 5 °C but was depressed at temperatures above this. The production of bacterial cell material in the waters of Drake Passage (62°30′S) is low compared with that in most other seawaters but amounted to 15–45% of the value for photosynthetic cell production. This activity is sustained by uptake of dissolved organic material. Although the overall metabolic activity of the bacterioplankton is heterotrophic we do not know the biochemical details. The bacterial flora of ocean waters is largely uncharacterized because so far most of the forms present have evaded isolation in culture. Analysis of bacterial 16S ribosomal genes from natural populations in the Sargasso Sea, on the Tropic of Cancer, has shown the presence of many undescribed and so far uncultured species and we may be sure that the same situation holds in polar waters. This is dramatically illustrated by the finding of DeLong *et al.* in 1994 that Antarctic waters are particularly rich in archaebacteria, a major group of prokaryotes once thought to be restricted to hypersaline, extremely hot, or anoxic habitats, but now known to occur in ordinary seawater. So far, these archaebacteria can only be shown to be present by measuring the binding of group-specific oligonucleotide probes to ribosomal RNA extracted from the mixed picoplankton assemblage. This technique showed them to be present to the extent of up to 34% of the prokaryotic biomass in samples taken in Arthur Harbour (64°46′S 64°04′W) in the late austral winter. They have also been found in Alaskan waters. The presence of such a substantial fraction of microorganisms with unknown biochemical activities points to our ignorance of what goes on at the lowest levels of biological activity in the sea. For viruses one can make a more informed guess. Electron microscopy of unpolluted seawaters shows the presence of viral particles in concentrations of around $10^8\,ml^{-1}$ and it seems likely that polar waters have a similar content. They attack bacteria and cyanobacteria (i.e. are bacterio- or cyanophages), and it seems likely that they have high multiplication and high decay rates with correspondingly high rates of lysis of host cells and release of soluble cell material into the water.

Bacteria and picophytoplankton are grazed by phagotrophic flagellates, some species of which combine this mode of nutrition with photosynthesis. They are ‘sloppy feeders’, releasing much dissolved organic matter from their prey in the course of eating it. It is these organisms that are chiefly responsible for the mineralization of organic matter to regenerate the inorganic forms of carbon, nitrogen, and phosphorus on which the phototrophs depend. Flagellates multiply rapidly and can keep pace with the growth of their food organisms. They themselves are preyed on by ciliates.

We have here a highly dynamic and closely knit community, based on photosynthesis by picophytoplankton. There is also high production of cell material by bacterioplankton at the expense of organic matter coming from release of soluble products of photosynthesis from healthy picophytoplankton, lysis brought about by viral infection, and sloppy feeding and excretion by the phagotrophs. Transfer of organic substrates and mineral nutrients by molecular diffusion is extremely rapid and efficient over the short distances, less than 100 μm, between cells and it does not seem that their growth can normally be limited by nutrient concentrations. Larger organisms are poor competitors for nutrients because of their low surface/volume ratios and the ultraplankton can thus form a self-sustaining community within which materials are recycled. Since the cells are tiny and sink at a negligible rate, loss of nutrients in sedimenting particles does not occur and the community is potentially capable of maintaining itself indefinitely in surface waters given adequate light. The population densities of the component species are set by the kinetics of their growth and trophic relationships rather than by input of materials from outside (Fig. 8.1). Changes of temperature have differential effects on these kinetics and in this seems to lie the explanation of a striking alteration in the composition of the picophytoplankton with latitude. In the North Atlantic, numbers of picoplanktonic cyanobacteria fall from $1.8 \times 10^5\,ml^{-1}$ at 38°00′N to $2.0 \times 10^3\,ml^{-1}$ at 58°32′N. They are nearly absent in Arctic Ocean water.

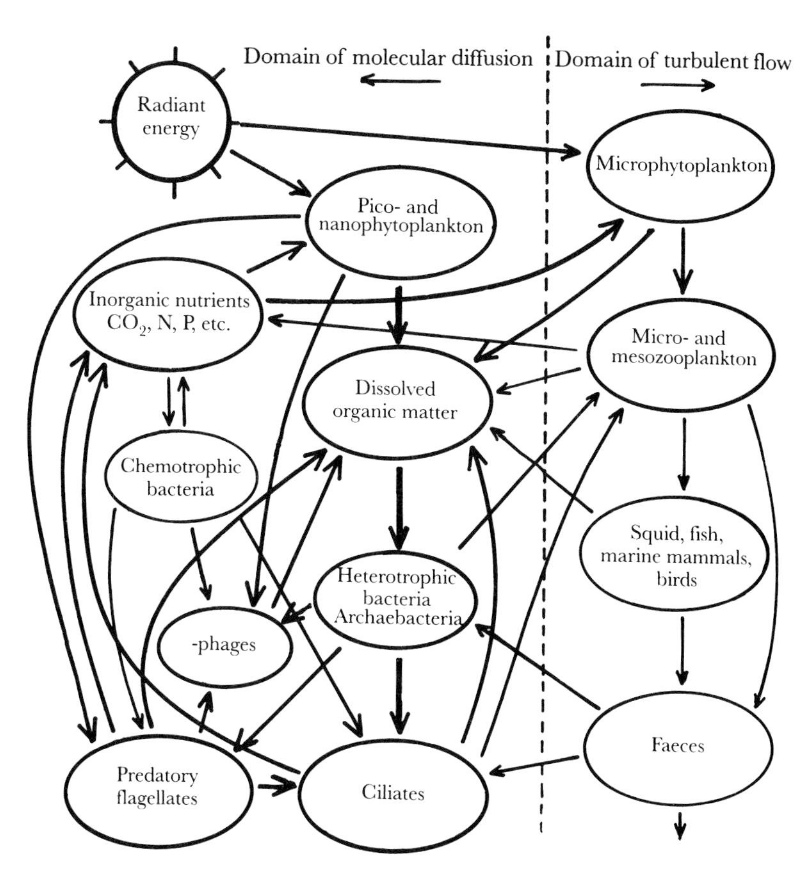

Fig. 8.1 Flows of energy and materials in the marine pelagic ecosystem showing the interrelations of the ultraplanktonic community with higher trophic levels.

South of Australia cyanobacterial numbers are correlated with temperature, but not with day length, varying from about 10 cells ml^{-1} at $-1\,°C$ near the Antarctic continent to about 10^4 cells ml^{-1} at $12\,°C$ further north. Eukaryotic picophytoplankton tend to increase in abundance with decreasing temperature, replacing the cyanobacteria as the primary producers. Heterotrophic picoplankton shows a less pronounced decrease with lower temperatures.

The extent to which the ultraplankton contributes to the sustenance of larger organisms is disputed. It was originally thought that the microbial loop is important in returning energy and materials *via* heterotrophic bacterioplankton to the classical pelagic food chain involving microphytoplankton and metazoans. It is not as simple as this. Picophytoplankton contribute as much cell production as the bacteria and it is now known that remineralization is carried out by flagellates to a greater extent than by bacteria. An experiment using radiocarbon-labelled substrate in temperate seawaters showed that in 13 days only 2% of the tracer moved from the ultraplankton into the larger organisms. In another study the predator–prey link between the phagotrophs and the bacteria was found to be tightly controlled through a chain of four trophic levels. Passage through this number of levels inevitably implies that little of the original production reaches the final level. There are suggestions that this may not be true in Antarctic waters (pp. 152, 191). There is evidence that ultraplankton is of importance in sustaining benthic suspension feeders in the winter (Clarke and Leakey 1996). Furthermore, appendicularians, macroplanktonic animals with an effective mechanism for filtering particles of bacterial size from seawater, are widely distributed and sometimes abundant in polar waters. They may provide a means of tapping the productivity of the ultraplankton for the ultimate benefit of fish, seabirds, and sea mammals.

8.2.2 The microphytoplankton

The life of microplankton is dominated by turbulent flow rather than by molecular diffusion. In oligotrophic waters it competes poorly for scarce nutrients with the picoplankton. Polar waters, however, usually contain plant nutrients in ample concentrations and microplankton is able to develop if other conditions are suitable. Because of its larger size microphytoplankton escapes the voracity of the rapidly multiplying phagotrophic flagellates. The planktonic metazoans which graze microphytoplankton take weeks to grow and reproduce so that there is a long lag between increase in their numbers and that of their prey. Population densities of microphytoplankton are thus less tightly controlled than those of picophytoplankton and variations are much greater, up to five orders of magnitude. The community is essentially opportunistic and non-equilibrium in nature. Whereas the ultraplankton seems fairly uniformly distributed in polar seas, the microplankton is patchy in abundance, its growth being largely controlled by water column stability.

8.2.2.1 The physiological ecology of polar phytoplankton

The general relations of photosynthetic rate to light conditions have been outlined in Chapter 2 (p. 44; Fogg and Thake 1987; Smith and Sakshaug in Smith 1990; Dring 1992). At depth, there is a compensation point at which gross photosynthetic assimilation just balances consumption in respiration and net photosynthesis is zero. Obviously, with continual change of irradiance arriving at the water surface, this is not a point for which experimental determination at any given time has much meaning but if mean rates over the daylight hours or over 24 hours are taken, the compensation point is a useful concept. If, for the algal population as a whole, the average amount of radiation received is less than the compensation value, then it will be unable to grow. It follows that growth of phytoplankton will only occur if the depth of mixing is less than a *critical depth*, related to the compensation point and determined by the prevailing irradiance and the transparency of the water. Conventionally, this is taken as the depth at which irradiance is 1%

of that at the surface but this is probably too high a value for polar waters. If the water column is stable so that a given cell remains at the same depth for a period commensurate with its generation time then adaptation can occur. Near the surface, cells become less susceptible to photoinhibition and photosynthesize at higher rates than they otherwise would. At depth, cells become shade-adapted and able to photosynthesize at higher rates than non-adapted cells (p. 46). On the other hand, if the water column is mixed, a cell may be carried between the surface and the compensation depth and its photosynthetic capacity will adjust to some intermediate, low level of irradiance. This usually means that all cells become shade-adapted (see Fig. 2.6). An index of shade adaptation is the assimilation number, p_m^B, the amount of carbon fixed per hour per unit amount of chlorophyll *a* at light saturation. Assimilation number is low, with a value of about one, in turbulent waters, in contrast to values of up to ten or more in phytoplankton adapted to high irradiance.

Normally, most of the photoinhibition near the water surface is caused by excess absorption of visible light. Additionally, there may be inhibition by ultraviolet (UV) radiation. In transparent oceanic waters about 1% of surface UV-B irradiance reaches 60 m depth, whereas in inshore polar waters significant amounts, say 2.5%, may reach 15 m depth and biological effects may be detected at 20 or 30 m. In well-mixed waters this may not be important since individual cells are near the surface for short times only and have ample time for repair in the shade of deeper water, so effects of UV may be overestimated. It is a different matter in stratified water, as in the marginal ice zone, where actively growing phytoplankton is held in a 10–20 m surface layer in its season of maximum growth – just when the 'ozone hole' is open. In the marginal ice zone of the Bellingshausen Sea, at around 64°S 72°W, in the austral spring of 1990 (when the edge of the 'ozone hole' touched the tip of South America) observations were made both inside and outside the area under the 'hole' (Smith *et al.* 1992). The sampling areas were selected on the basis of daily transmissions of ozone concentrations from the NASA Nimbus 7 satellite. The ratio of UV-B to total irradiance at the sea surface increased under the 'hole' and there was an estimated minimum of 6–12% reduction in primary productivity. Although this should be viewed in relation to a ±25% year-to-year variation which usually occurs, this is an appreciable effect. Numbers of *Phaeocystis* in the surface waters were positively correlated to ozone concentrations, the decline in numbers being seemingly due to cell lysis caused by increased exposure to UV-B. Diatoms were not affected in this way. Two important ecological consequences of such effects can be envisaged, although so far there is no evidence that either has occurred. First, there may be selection for UV-B resistant species. *Phaeocystis* produces higher concentrations of UV absorbing compounds than do diatoms, which show great variability between species in this respect. However, ability to survive exposure to UV is not correlated with production of these compounds and some diatoms, using other mechanisms to mitigate UV damage, survive irradiances which are lethal to Antarctic *Phaeocystis*. One might expect, therefore, that in the course of time ozone depletion will lead to replacement of *Phaeocystis* by resistant diatoms. This might be a 'good thing' in that the nutritional value of *Phaeocystis* for zooplankton is low whereas that of diatoms is better and might provide for greater production of fish. The second possibility is the rather surprising one that UV-B damage to phytoplankton might have effects on climate. Some species, e.g. *Phaeocystis*, release the volatile compound, dimethyl sulphide, a derivative of the osmotically compatible solute (p. 36) dimethylsulphoniopropionate. It is estimated that as much as 10% of the total global flux of this substance into the atmosphere comes from the Southern Ocean. In the atmosphere dimethyl sulphide is oxidized to sulphate to form aerosol particles which act as cloud condensation nuclei. The albedo of clouds strongly influences global climate and is itself determined by the concentration of these nuclei. Thus any alteration in the abundance or distribution of *Phaeocystis* induced by UV damage could have dramatic consequences (Marchant in Hempel 1994).

Polar phytoplankton species do not possess adaptive mechanisms to enable them to photosynthesize and grow at faster rates at low temperatures than similar forms from temperate waters would do under the same conditions. Temperature is an important factor controlling metabolic activity of phytoplankton in the Arctic and Antarctic. Indeed, it appears to influence a process which is considered to be normally independent of temperature, (i.e. light-limited photosynthesis, see p. 45). In the Scotia Sea and Bransfield Strait in an area centred on 61°30′S 57°0′W, assimilation numbers of phytoplankton at saturating light levels, P^B_m, and the slopes of the light-limited region of the photosynthesis versus irradiance curves, α^B, have been found to be lower than in algae at lower latitudes. P^B_m, as expected, increased with rise in temperature, having a Q_{10} of *c.* 4.2 between −1.5 and +2 °C, whereas, unexpectedly, α^B also increased, with a Q_{10} of *c.* 2.6 between −1.5 and 5 °C. Above 5 °C there were no increases with rise in temperature in either rate. It seems that at extremely low temperatures some temperature-dependent reaction becomes rate-limiting for the photochemical reactions (Tilzer *et al.* 1986). Psychrophilic marine diatoms are capable of some photosynthesis at −8 °C.

Deep mixing may carry phytoplankton below the critical depth and there is also the darkness of the winter months. It seems most probable that polar algae survive these deprivations by reducing basal respiration to a minimum so that cellular reserves suffice to carry them through. Some planktonic algae are capable of heterotrophic nutrition and for those flagellates which are capable of both photosynthesis and phagotrophy there is no problem. By and large, however, it seems that the utilization of dissolved organic matter to enable survival during periods of darkness does not occur. That some diatoms can survive long periods of darkness is not in doubt. Cultures of sea ice algae from both Arctic and Antarctic kept in the dark at −2 °C were still viable after 12 months. Survival was increased if darkness was imposed gradually as would happen under natural conditions (Kirst and Wiencke 1995).

Psychrophilic algae have a low temperature optimum, usually below 15 °C, for net photosynthesis. This has been explained by supposing the Q_{10} for gross photosynthesis to be less than that for respiration which thus progressively overtakes the other as temperature is raised. In fact, the Q_{10}s for photosynthesis in entire samples of Antarctic plankton are about the same as those for respiration. However, it has to be remembered that in such samples the respiration of both phototrophs and heterotrophs is included in the overall determinations. The problem thus remains unresolved (Kirst and Wiencke 1995).

The products of photosynthesis can only be elaborated into cell material and result in growth if the essential elements and, in certain cases, specific organic moieties, are available. Some of these, such as carbon, hydrogen, oxygen, potassium, calcium, and sulphur, are in ample supply in seawater but other major nutrients, notably nitrogen, phosphorus, and silicon (necessary for diatoms and silicoflagellates but not for other kinds of algae) are often present in low concentrations which commonly limit growth. The Southern Ocean receives a constant supply of nutrient-rich water by upwelling at the Antarctic Divergence (p. 12) and has concentrations of these elements about twice as high as those of the most fertile areas elsewhere in the world ocean, so they are rarely limiting. There are, however, indications that silicate may sometimes limit growth of diatoms immediately south of the Polar Front (Knox 1994; Kirst and Wiencke 1995). The situation in the Arctic is more complex, with no major input as in the Antarctic but lesser contributions from the Siberian and Alaskan rivers, from the Atlantic via the Norwegian Current, from the Pacific via the Bering Strait, and from regeneration in the shelf regions. Nutrient concentrations are low in the surface waters of the Arctic Ocean but increase at the halocline (p. 11) at depths where utilization by phytoplankton is minimal. Elsewhere, nutrient concentrations are high where upwelling of deeper water occurs at fronts, as in the Labrador Sea. High production in the western Bering Strait region is supported by a cross-shelf flow of nutrient-rich water from the Bering Sea continental slope. However, nutrient depletion is common in Arctic waters.

Micronutrients include trace metals and organic growth factors such as biotin, thiamine (vitamin B_1), and cyanocobalamin (vitamin B_{12}). Vitamins are not required by all species and are unlikely to limit total standing crop but may determine its species composition. Recently, attention has focussed on iron as the trace element most likely to affect phytoplankton abundance. Iron is a constituent of several vital enzymes and is fourth in order of abundance of elements in the earth's crust. River waters contain high concentrations of iron in the form of complexes with organic matter and coastal waters are correspondingly rich in this element. The chemistry of iron in oceanic waters is complicated but ferric hydroxide, the main inorganic form, is sparingly soluble and, being readily adsorbed on particulate matter is removed by precipitation. Reported measurements of iron concentrations are suspect because it has not always been realized how easily samples become contaminated. However, oceanic waters appear generally to have low concentrations, their principal supply coming as fall-out of dust derived from the land. Arctic seas get sufficient supplies from exposures of rock and drylands in their vicinity. The Antarctic is much less favourably situated since dry exposed land lies far to its north and out of the path of the prevailing westerly winds. The Southern Ocean may consequently be deficient in micronutrients such as iron. The evidence for iron limitation in Antarctic waters is conflicting. Carefully performed experiments in which various nutrients were added to seawater samples taken from the southwest Atlantic sector have failed to show any stimulation of phytoplankton by iron (Hayes *et al.* 1984). On the other hand, the waters of the Polar Front, which produce phytoplankton blooms, are rich in iron whereas iron-poor waters further south are much less productive (De Baar *et al.* 1995). Sediment cores from the Atlantic sector of the Southern Ocean show periods of increased accumulation of biogenic detritus during glacial periods coinciding with increases of up to fivefold in iron content. It appears that greater transport of dust from Patagonian deserts during the last glacial maximum led to conspicuous stimulation of phytoplankton by increasing iron supply (Kumar *et al.* 1995). Probably, there are differences between the ultraplankton and the microplankton in respect to iron. There is extremely efficient uptake from trace concentrations combined with rapid recycling within the ultraplanktonic community (p. 184). Some of the microbial groups involved – bacteria, cyanobacteria, and chrysophycean flagellates – produce *siderophores*, powerful complexing agents which bind trace metals, such as iron, in a form which can only be utilized by the organism which produces them. Even if the microplankton is iron-limited in the Southern Ocean, the ultraplankton, perhaps, is not. There has been the idea of fertilizing Antarctic waters with iron as a step towards mitigating the 'greenhouse effect' by promoting carbon dioxide uptake by phytoplankton. One would guess, however, that any addition of iron would only be short term in its effects.

The characteristic seasonal cycle of microphytoplankton in polar waters is represented by a bell-shaped curve, rising from low levels in spring to a peak around midsummer and declining again to a low level by early autumn. This contrasts with the curve for temperate waters which has peaks in spring and autumn with a trough in the summer. Thus, in polar waters phytoplankton growth follows the seasonal cycle of radiation fairly closely, the slow development of herbivores at low temperatures delaying their impact until their prey is already in decline as a result of diminishing radiation and nutrient supplies and increasing turbulence. In temperate waters, the herbivores develop more quickly and their grazing, combined with exhaustion of nutrients, reduces the standing stock of phytoplankton by midsummer. Nutrient regeneration follows and temperature, water column stability, and radiation allow a second peak in autumn (Fogg and Thake 1987). Ultraplankton in Antarctic waters has a seasonal periodicity similar to that of the microplankton but the curve of standing crop is flatter and the peak occurs later in the summer. The standing crop in the winter is about five times that of the microplankton – a point which may be of great importance for both zooplankton and zoobenthos (Clarke and Leakey 1996).

Phytoplankton growth in polar seas is patchy, especially in the Antarctic where vast areas of oceanic desert contrast with limited regions of intense productivity. In the Arctic, the causes of patchiness are not far to seek. Over most of the ice-free area in summer there is stratification with a pycnocline at about 25 m and the rather low nutrient concentrations in the waters above it are soon depleted. Production then depends on regeneration within the surface layer and increase can only occur if disruption of the pycnocline introduces nutrient-richer water from below. This can be brought about by storms; shear forces between currents, banks, and islands; vertical motions induced by upwelling or eddies; or tidally induced mixing in shallow waters. An example is a particularly rich area in the western Bering Strait where in a three-month summer season most of an annual yield of $324\,g\,C\,m^{-2}$ is produced. Growth during the winter is inhibited by lack of light, deep mixing, and ice cover, and nutrients accumulate. When the ice retreats and the mixed layer depth decreases conditions become favourable for phytoplankton growth. This growth is maintained by upwelling of nutrient-rich water along the edge of the continental shelf, which is then carried across the shallow shelf by a current flowing northwards along the western coast of the Bering Strait. This produces a situation rather like that in a laboratory continuous culture, in which nutrients are replaced as rapidly as they are taken up. A conservative estimate of the total yield of primary production in Arctic waters is 0.21×10^9 tonnes $C\,yr^{-1}$ (including production by ice biota). This is about 0.6% of the total for the world ocean. This implies, since the area of sea is about 4% of the total, that primary production is less efficient than in other regions.

In the Southern Ocean, the reasons for patchiness are more obscure. Conditions of light and temperature are similar all round the continent and everywhere nutrient concentrations are high. The prevalent westerly winds generally mix the surface layers to an extent sufficient to prevent stratification. Where ice melt produces local stratification plankton production is intense. The importance of stability of the water column is well illustrated by studies in inshore waters around Signy Island. Phytoplankton begins to increase as fast ice disappears and reaches a peak a month or so later. There is then an abrupt decline although water temperature and light are still relatively high, ice is not yet re-forming, and nutrients, as exemplified by nitrate, are not exhausted (Fig. 8.2). The crop is not grazed to any significant

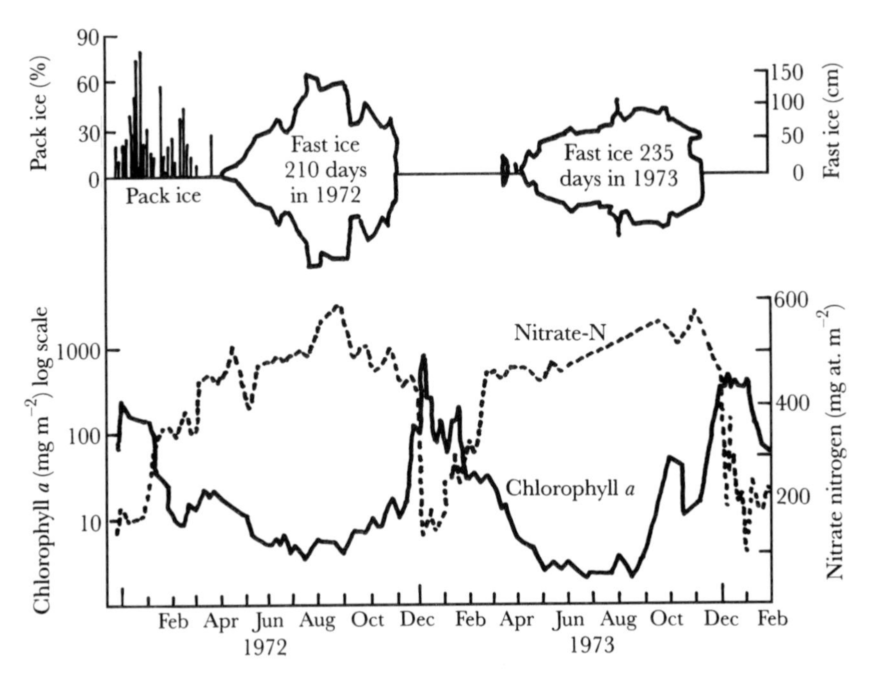

Fig. 8.2 Observations over two years of amounts of ice, phytoplankton chlorophyll, and nitrate concentration in the water column (16.5 m mean depth) in Borge Bay, Signy Island. (Data courtesy of Dr T. M. Whitaker.)

extent and the decline coincides with high winds, presumably extending mixing below the critical depth. It seems possible, but not yet convincingly proven, that the low standing stock of phytoplankton generally found in the Southern Ocean is likewise due to turbulent mixing. Growth in these circumstances can only take place where hydrographic features, such as eddies, maintain populations in the photic zone. The upward vertical velocity in an anticyclonic eddy is probably around 1 m day^{-1} so this would offset to some extent the settling of microplanktonic cells, the velocity of which is in the range 1–10 m day^{-1}. This idea is supported by a correlation of high phytoplankton densities with upwelling, as indicated by low surface temperature, or with features of the seabed, such as shelf breaks, submarine mountains or islands, which produce eddies (Fig. 8.3). It is unlikely that these effects are due to nutrient enrichment from deep water, although it could be that supply of iron or other micronutrients is sometimes important. Although grazing by zooplankton can dramatically reduce phytoplankton concentrations locally this does not provide an explanation for low levels over wide areas of the Southern Ocean.

Where phytoplankton is densest, values of 40 or more mg chlorophyll *a* m^{-3} with carbon fixation in excess of 2 g m^{-2} day^{-1} are reported. These are comparable with values from the most fertile areas in temperate regions. On

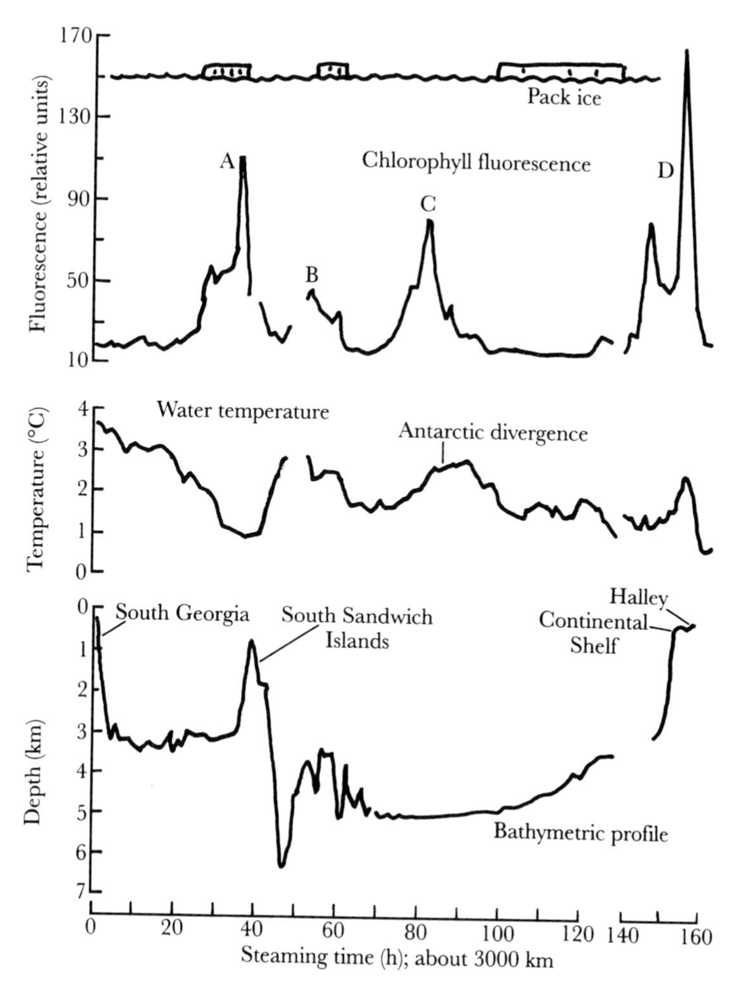

Fig. 8.3 Patchiness of phytoplankton in the Southern Ocean; observations of chlorophyll fluorescence, surface water temperature, and depth made during a cruise across the Weddell Sea from South Georgia to Halley Station. The chlorophyll peak A is an example of the 'island effect'; B, of phytoplankton growth in the marginal ice zone; C, perhaps also in the marginal ice zone and/or associated with the Antarctic Divergence at around 65°S; and D, in inshore waters of the Brunt Ice Shelf. (Data of Hayes *et al.* 1984.)

the other hand, in the open ocean values are usually around 0.5 mg chlorophyll *a* m^{-3} and 0.2 g C m^{-2} day^{-1}. Given this large variation and the paucity of data – which usually has to be obtained opportunistically from ships in passage, making it difficult to distinguish between seasonal and geographic variations – any calculation of the total production can only be tentative. Estimates are that an area of 32×10^6 km^2 has a primary production of 10^9 tonnes C per annum, which suggests that overall, Antarctic waters are less than half as efficient as other parts of the world's oceans.

8.2.2.2 The Arctic phytoplankton

In terms both of numbers of species and biomass diatoms predominate, dinoflagellates, other flagellates, and green algae being present in much smaller proportions. The species are not the same in different sea areas (Guillard and Kilham 1977). In the Barents Sea the important diatoms include *Chaetoceros diadema*, *Corethron criophilum*, *Skeletonema costatum*, and *Rhizosolenia styliformis*. Among the dinoflagellates are *Peridinium depressum* and *Ceratium longipes*, and of the green algae only *Halosphaera viridis* is widely distributed. In small bays and inlets there may be mass development of *Phaeocystis* sp. in spring. Some warmer water forms (e.g. *R. styliformis* and *Ceratium tripos*) are brought in by the North Atlantic Current. The Bering Sea divides into two distinct areas, cold water plankton being found on the western side and warmer water forms on the eastern. In the cold water the diatoms include *Thalassiosira nordenskiöldi*, *Chaetoceros socialis*, and *Eucampia groenlandica*, whereas on the other side are *Thalassiosira japonica*, *Rhizosolenia alata*, and *Ditylum brightwellii* with the dinoflagellate *Peridinium excentricum*. This distribution is related to currents. The Laptev Sea is different again, having the strong influence of the Siberian River Lena. The common forms in the offshore, highly diluted, region include brackish–freshwater species of cyanobacteria, *Aphanizomenon flos-aquae*, and *Anabaena* spp., and the freshwater diatoms, *Melosira italica* and *Asterionella gracillima*, with marine forms, *Thalassiosira baltica*, *Chaetoceros gracile*, and *Peridinium breve*, providing only about 5% of the population. There is no great degree of endemism amongst Arctic phytoplankton.

8.2.2.3 The Antarctic phytoplankton

As might be expected, the circumpolar circulation and sharp delineation by the Polar Front (p. 12) combine to given the Southern Ocean a relatively uniform plankton flora (Guillard and Kilham 1977). As in the Arctic the most prominent forms are diatoms (Fig. 8.4). Silicoflagellates are also abundant but dinoflagellates less so although, their cells being larger, they may sometimes exceed the diatoms in biomass. The predominance of diatoms is to be expected since they sink rapidly and hence are dependent on turbulence to maintain them in the photic zone. The larger flagellates, on the other hand, lose the biological advantage of being able to move to the optimum depth in the water column if there is vigorous mixing. There is no well-defined seasonal succession of species. The number of species thought to be bipolar has diminished as taxonomic knowledge has advanced. The supposedly characteristic Antarctic species *Thalassiosira antarctica* appears to be bipolar since it has been found in the northern hemisphere but not in latitudes between 58°S and 58°N. It may yet be found in intermediate locations or, quite possibly, it may not be genetically isolated from other *Thalassiosira* spp. from which, in fact, it is distinguished only with difficulty. Some species appear to be cosmopolitan but the *Phaeocystis*, now distinguished as *P. antarctica*, which is abundant in the Southern Ocean, has been

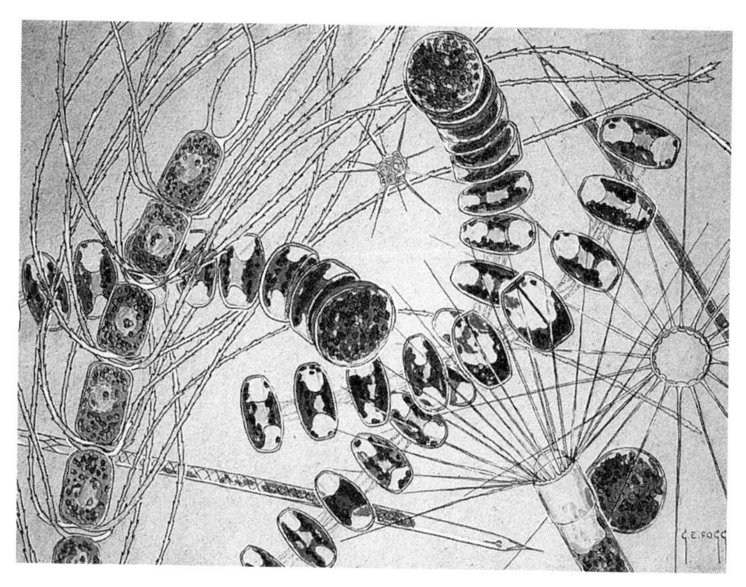

Fig. 8.4 Microphytoplankton from the north Weddell Sea, February 1966. The algae illustrated are mainly diatoms: *Corethron criophilum* (bottom right-hand corner, straight radiating spines), *Chaetoceros criophilum* (chain of cells with long barbed spines), *Thalassiosira antarctica* (chains of pillbox-like cells), and *Rhizosolenia* sp. (long needle-like cells), but the solitary star-shaped organism is a silicoflagellate, *Dichtyocha speculum*. (×130; watercolour drawing, G. E. Fogg.)

shown by sequence data from 18S small subunit ribosomal DNA to be genetically distinct from *P. pouchetii*, the northern cold water form, and from *P. globosa*, the warm water species from which the other two seem to have evolved (Medlin *et al.* 1994). Endemism is high among the Antarctic microphytoplankton, 80–85% of dinoflagellates and around 37% of diatom species. However, at generic level most of the important taxa in the Antarctic phytoplankton are cosmopolitan but the two monospecific genera, *Charcotia* and *Micropodiscus*, seem to be endemic.

8.2.3 The zooplankton

Nanozooplankton has been discussed under the heading of ultraplankton (p. 182). Larger forms, which are nearly all metazoans, are best dealt with together although they range in size from microplankton (20–200 μm), through mesozooplankton (0.2–20 mm), to macrozooplankton (2–20 cm). Most of these, with the exception of appendicularians (p. 186), feed on prey usually about two orders of magnitude smaller than themselves, which they either filter out or seize as individuals while actively swimming. They may be herbivorous, feeding on microphytoplankton, omnivorous, or carnivorous, feeding on other zooplankton. Most have life cycles with several developmental stages which are controlled by low temperatures and the brief abundance of food. Growth when food is available is fast but overall rates are slow and lifespans extended, because of long periods of zero or negative growth when basic metabolism is maintained at the expense of reserves. Increased incidence of UV radiation may affect zooplankton indirectly by altering the amounts and kinds of food organisms. There are also deleterious direct effects but one supposes that the larger zooplankton may avoid these by vertical migrations.

In both hemispheres the chief groups represented are; radiolarians, coelenterates, rotifers, chaetognaths, copepods, ostracods, euphausiids, amphipods, mysids, pteropods, and appendicularians. The species are, however, different (Zenkevitch 1963; Smith and Schack-Scheil in Smith 1990; Knox 1994).

8.2.3.1 The Arctic zooplankton

Zooplankton abundance broadly reflects phytoplankton abundance (Zenkevitch 1963). Thus, the central Arctic Ocean has sparse zooplankton with biomass around 1–3 $mg\,m^{-3}$, about a thousandth of that in the Bering Sea. The main input of water into the Arctic basin being from the North Atlantic one finds Atlantic species, such as *Calanus finmarchicus* and *Metridia lucens*, but the most frequent copepods are *C. hyperboreus*, *C. glacialis*, and *M. longa*. Long-term studies of these have been made using ice islands as research platforms in the centre portion of the basin. The copepods predominate near the surface in the summer. The life cycle of *C. hyperboreus* seems to take three years; one year is required for development from egg to copepodid stages II and III, the second for stages IV and V, and the third for adulthood and spawning. Copepodids stay below 300 m in winter, migrating to the surface 100 m in summer. Gravid females descend slowly from the surface down to 300 m in spring, while males, which are fewer in number, go below 400 m.

The marginal seas of the Arctic basin have a similar pelagic fauna to that of the central parts with the addition of neritic species dependent on the temperature and salinity conditions arising from the inflow of the rivers. The Kara Sea shows particularly distinct zones with faunas indicating the origins of the waters – in the north from the Arctic basin, in the south from the Barents Sea, and an intermediate zone of brackish water with freshwater species occurs in waters from the rivers Ob and Yenisei.

The high primary productivity of the western Bering Strait (p. 191) is matched by secondary production (Sambretto *et al.* 1984). The zooplankton species are characteristic of the North Pacific. The large copepods, *Neocalanus plumchrus* (up to 5 mm length) and *N. cristatus* (up to 10 mm length), make up 70–80% of the zooplankton biomass in summer. They overwinter at depth. Eggs are produced from autumn to spring and the young rise to the surface from early spring onwards. In the Bering Sea much of the annual growth takes place during the dense spring phytoplankton bloom. In the sub-Arctic Pacific further south there is ample phytoplankton on which to graze over most of the year since a pycnocline keeps the mixed layer shallow and some photosynthesis is possible during the winter. The copepods are present in sufficient numbers to keep the standing stock of phytoplankton at a low level even in summer and if it becomes inadequate they turn to predation on the abundant microzooplankton.

The predominant form in the Barents and Greenland seas is *Calanus* but there has been confusion because this is an area of sympatry between two closely related species, *C. finmarchicus* and *C. glacialis*, the former being found in temperate as well as Arctic seas whereas the latter is endemic to the Arctic. There is morphological similarity between the two, although *C. glacialis* is the larger, but if collected alive their swimming patterns and behaviour are different enough to distinguish them. The situation is further complicated because *C. glacialis* has a variable life cycle, lasting two years with spawning in May and June in the Barents Sea but a one-year cycle with spawning in December off western Greenland. In the southern part of the Barents Sea the development of *Calanus* spp. occurs in early summer but, north of 72 or 73°N, in the autumn. The biomass may attain as much as 300 mg fresh weight m^{-3} of which 90% or more is *Calanus*. It is the key organism in the food web. Large numbers of herring, haddock, and the fry of various other fish feed on it in competition with jellyfish, and ctenophores, which may reach as many as 123 individuals m^{-3}.

8.2.3.2 The Antarctic zooplankton

As with the phytoplankton, the zooplankton assemblage of the Antarctic has a generally circumpolar distribution. There is, however, latitudinal zonation as seen, for example, in the distribution of euphausiids (Fig. 8.5). A wide variety of species belonging to several groups is found but this is often overlooked because of the dominance of one species which is generally regarded as the key organism in the food web and has attracted attention as a possible economic resource. This is *Euphausia superba*, the Antarctic krill – 'krill' being derived from the Norwegian *kril*, originally used by whalers to denote small fish and thence transferred to cover crustaceans eaten by baleen whales. *E. superba* is a shrimp-like macroplanktonic form with a length of up to 60 mm and a fresh weight of about a gram. It is bigger than any corresponding Arctic organism. It is stretching a point to describe it as 'planktonic', implying that it drifts at the mercy of tides and currents, since it can swim at a speed of around 13 cm s^{-1} and krill swarms have been seen to move against prevailing currents. This mobility and the often patchy distribution make for great difficulties in precise studies of krill in their natural environment. Modern nets can yield reasonably quantitative results with micro- and mesozooplankton but not when, as with krill, the organism can detect the approach of a net and take rapid avoiding action. Large trawls without attachments directly in front of the net give the best results but even these are effectively avoided during the day. A sonic technique for

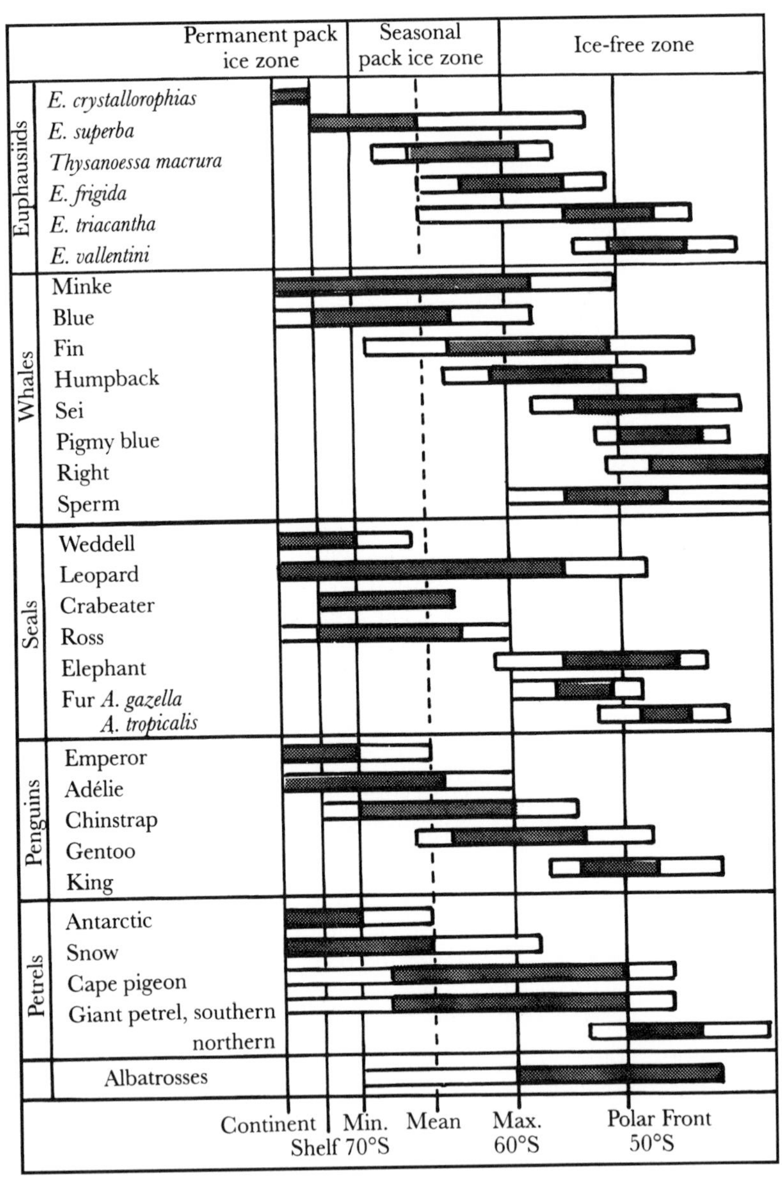

Fig. 8.5 Zones occupied by selected species of euphausiids, marine mammals, and birds, from the Antarctic continent northwards. Each species has a circumpolar distribution and the average latitudinal range is given, with the shaded areas indicating the greatest densities. (Redrawn after Laws 1977 by courtesy of the author.)

sampling is proving more useful. This is a modification of echo-sounding, by which surfaces reflecting acoustic signals can be detected by the echoes received from them. The time elapsing between dispatch of the signal and receipt of the echo is a measure of the distance of the reflecting surface and its strength is a measure of the amount of material in it. Integrated data from an echo-sounder operating at an appropriate frequency give an mean volume backscattering strength from which the density of a krill swarm can be estimated, given a target strength determined experimentally (Fig. 8.6). However, krill in the surface layer, above the echo-sounder transducer, are not sampled and the technique functions poorly in pack ice.

Euphausia superba was studied intensively by the British *Discovery Investigations* between 1925 and 1939 (Hardy 1967), and by BIOMASS, an international programme of *Biological Investigations of Marine Antarctic Systems and Stocks*, between 1977 and 1991 (El-Sayed 1994). Nevertheless, knowledge of the life cycle and distribution of krill is still incomplete. It occurs in surface waters south of the Polar Front, showing most activity at −1.5 °C and much less at 4 °C. In this it contrasts with its ecological counterpart north of the Polar Front, the lobster krill *Munida gregaria*, which is most active at 8–10 °C and killed if transferred to 0 °C. *Euphausia superba* is found down to 200 m, but mostly between 20 and 100 m. It typically undertakes daily vertical migrations, rising to the surface at night and sinking to concentrate in swarms during the day evidently in response, although in no simple manner, to food levels and irradiance. The swarms are usually dense and often large, measuring up to 500 m or more horizontally. Occasionally, they are extremely large as, for example, a 'superswarm' found in March 1981 north of Elephant Island (*c.* 61°S 55°W), which acoustic observations showed as extending over 450 km^2 with a biomass of 2.1×10^6 tonnes. The movement of individuals in a swarm is co-ordinated and synchronized, the whole changing shape in an amoeboid fashion; this provides an entrancing sight

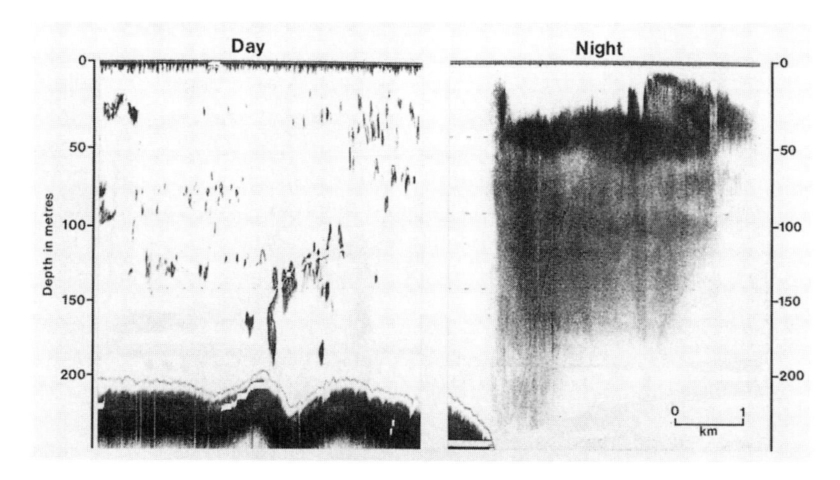

Fig. 8.6 Echo-charts showing differences in krill behaviour; small, compact, swarms, mostly at depth, were present during the day but during the night these dispersed into an extensive low density layer near the surface. (Courtesy of Dr I. Everson and the British Antarctic Survey.)

for divers, especially at night because krill has a blue luminescence. Swarms often differ in composition in terms of size of individuals, stage of maturity, sex ratio, and feeding state. Feeding tends to be more active when the krill is dispersed. The food is mainly phytoplankton but small zooplankton, even their own larval stages, may be taken. Prey is filtered out in a basket formed by the animal's fringed thoracic limbs. *Euphausia superba* does not have great reserves of lipid to carry it through the winter but can withstand starvation for long periods by utilizing its body protein. It may frequent the undersurface of pack ice (p. 166), feeding on attached algae by raking them off with its thoracic appendages. When the ice melts and algae are released, normal foraging behaviour, with increased swimming rates, is induced. The ice provides for overwintering adults and a nursery for the larvae but the extent to which krill populations as a whole take advantage of this is not yet established. Krill also overwinters near the sea bottom in shallow bays, feeding on detritus.

Growth and longevity are difficult to establish from studies of natural populations and laboratory observations may not be reliable guides to what happens in the sea. Biochemical features indicate that *E. superba* has a lifespan of 5–8 years. Spawning may take place three or more times during the life of a female, up to 8000 eggs being released at a time. This happens at depths around 100 m and the eggs immediately sink at a rate of 150–250 m day^{-1} and hatch at around 1000 m. Whereas adults are carried northeast towards the Polar Front in Antarctic Surface Water, eggs and young larvae are carried back south in the Circumpolar Deep Water (Fig. 1.5). As the larvae develop they ascend slowly to the surface, so that different stages are found stratified in order with the young adults eventually arriving back once more at the surface. In this way, *E. superba* is able to maintain itself over the latitudinal extent of the open Southern Ocean.

The distribution of krill, however, is not uniform. It is particularly abundant, for example, in the Scotia Sea, around South Georgia, and north of the Ross Sea. This pattern seems generally explicable in terms of the transport of larvae and young adults by currents and their collection in gyres and eddies. There seems to be only one genetic population of *E. superba*. Recent electrophoretic studies on enzyme proteins have not confirmed an early report that there are at least two discrete populations.

The total standing stock of *E. superba* has been variously estimated as between 14 and 1000 million tonnes – clearly a large amount by global standards and probably exceeding the total biomass of the human race. Because krill is of ecological importance and a possible economic resource (p. 221), it is necessary to know the rate at which this standing stock produces new biomass. Given the inadequacy of knowledge of its population dynamics and survival during the winter, this cannot yet be done satisfactorily from basic principles. For an empirical estimate one needs to know the biomass of the stock at the beginning and end of some set period, say a year, and the allowances which must be made for losses by catabolism, mortality from predation and other causes, and emigration or immigration from or into the area. Again, these are difficult to determine and indirect methods have to be used. Perhaps the best estimate comes from data on consumption by major predators. Now that whale numbers have been reduced they take between 34–42 million tonnes per year. This is some 150 million tonnes less than they consumed before exploitation but it is fallacious to think that there is now a surplus there for the taking. Populations of other predators have probably increased in recent years as they benefit from the reduced competition. Consumption by seals, squid, fish, and penguins is estimated at 130, 30–50, 10–20, and 25–50 million tonnes per year, respectively, which, including that of whales, gives a total of between 235 and 290 million tonnes per year. This can be equated, roughly, with the total annual production (Miller and Hampton 1989; Knox 1994; Nicol in Hempel 1994).

Why should *E. superba* be so successful? It has been described as 'a creature of great agility, powers of locomotion, purposeful intent and not a little awareness'. Apart from this, its fecundity and capacity to re-establish its numbers in the face of staggering inroads by its many predators, its freedom from parasites, and its versatility and effectiveness in exploiting both the open ocean and the sheltered habitat of the pack ice must all contribute.

Other important zooplankters in the Southern Ocean are chaetognaths, such as *Sagitta gazellae*, copepods, such as *Calanoides acutus*, and euphausiids such as *Thysanoessa macrura*. Like *E. superba* these often appear in swarms. Reproduction is usually timed to coincide with the seasonal pulse of phytoplankton production. The salp, *Salpa thompsoni*, a barrel-shaped organism which filters phytoplankton as it pumps the water through itself, shows remarkable adaptation to the situation. During the winter it exists in a solitary form but this, in the spring, buds off, asexually, large numbers of aggregate forms. Each solitary organism can produce about 800 aggregates when triggered, presumably, by the phytoplankton bloom, so that dense swarms are built up during the summer. The mean biomass of zooplankton other than *E. superba* is estimated as 10–50 mg wet weight m^{-3}, going up to 300 mg wet weight m^{-3} at the Polar Front. This is probably about the same biomass as for *E. superba*. A feature of Antarctic meso- and macro-zooplankton is that its biomass in the upper 1000 m scarcely fluctuates over the year, the bulk of the stock consisting of a few large species which overwinter in the deeper water. It may be that these feed on the ultraplanktonic community, which also tends to a steady standing stock over the year, as has been suggested for benthic filter-feeders (p. 150). These zooplankters are not a major food source for large predators such as prey on *E. superba* but they are eaten by the amphipod, *Parathemisto gaudichaudii*. This has a life cycle of one year geared to produce juveniles at around the time when herbivores are producing theirs. It has marked diurnal vertical migration, collecting at the surface at night, but it does not swarm in the way that herbivores do.

8.3 The upper trophic levels

The larger animals – squid, fish, birds, seals, and whales – the top carnivores in the open seas of the polar regions, are again mainly of different species north and south. However, their roles in the ecosystem are generally similar and Arctic and Antarctic may be dealt with together (Ainley and DeMaster in Smith 1990).

8.3.1 Squid

Squid are abundant and ecologically important in both polar regions – they are a major food of the male sperm whales, which migrate into both polar seas, and there are important fisheries for some species around Alaska and in sub-Antarctic waters. Nevertheless, we know little about them. Squids are fast-swimming, with well-developed nervous systems, and are difficult to catch in ordinary nets. Both in species and size there is little correspondence between those caught in research nets and those whose remains are found in predators such a sperm whales (Rodhouse in Kerry and Hempel 1990). The food of squid evidently consists largely of euphausiids, other crustaceans, smaller cephalopods, and fish. They themselves are important in the diets of killer whales, seals, birds, such as the emperor penguin and albatross, and larger cephalopods. An estimate, admittedly rough, is that in the Southern Ocean the consumption of squid by predators is around 30×10^6 tonnes per annum. The standing stock necessary to support this may well be in excess of 100×10^6 tonnes (Knox 1994).

8.3.2 Fish

Fish belonging to a variety of groups are found in Arctic waters but notothenioides, which are dominant in the Antarctic, are notably absent. In the high Arctic, the Arctic cod is a key species. It feeds on euphausiids, copepods, and pteropods, and, in coastal waters, benthic amphipods. In turn it is preyed on by the ringed seal, white whale, Arctic tern, glaucus gull, black guillemot, and other seabirds. Where there is freshwater input around the Arctic basin many species of Salmonidae are abundant, particularly off the Alaskan and eastern Siberian coasts. Arctic char is one of these, abundant in streams but fattening in the sea in its fourth and fifth years before it spawns in freshwater. In subpolar regions, capelin (*Mallotus villosus*) winters in deep waters, rising to form huge surface shoals in summer. Eel pouts (Zoarcidae), polar cod (*Arctogadus glacialis*), and polar halibut (*Reinhardtius hippoglossoides*) occur in deeper waters. In the low Arctic the fish faunas are derived from adjacent temperate seas; capelin, herring (*Clupea harengus*), and sandlance (*Ammodytes hexapterus*) in the northeast Atlantic; walleye pollock (*Theragra chalcogramma*), and herring (*Clupea pallasi*) in the north Pacific (Zenkevitch 1963; Ainley and DeMaster in Smith 1990). These and other cold water species provide the basis for commercial fisheries.

The Antarctic fish fauna is more homogeneous than that of the Arctic. 65 per cent of nototheniid species have a circumantarctic distribution. Nevertheless, zoogeographical units can be distinguished. There are differences at the generic level between the fish in the waters around East and West Antarctica and there are endemic species associated with island groups. Diversity is low; a sea area which amounts to 10% of the world's oceans contains only about 1% of the fish fauna and in shelf and slope waters 55% of the species fall into one group, the Notothenioidei. This has not always been so. Fossils from Seymour Island (64°17′S 56°46′W) show that in the late Eocene, warm seas over extensive shallow shelves supported a diverse and cosmopolitan assemblage of fishes, few of which are represented taxonomically in the region at the present day. One must assume that the establishment of the Circumantarctic Current some 25 million years ago and the subsequent onset of frigid conditions caused their extinction. No fossil record of notothenioides has yet been discovered and the group seems to have appeared in Antarctic waters out of the blue. That 97% of species and 85% of genera are endemic points to their evolution in isolation. Possibly the notothenioides, originally a benthic nearshore group, survived or possibly they were most successful in invading across the Polar Front. This front seems to be a barrier, preventing invasion by temperate water fish, but it may not be simply the temperature drop across this front which is the operative factor – after all, in the Arctic temperate species are able to establish themselves quite successfully in seasonally cold waters. Survival in constantly cold water requires evolutionary adaptation rather than temporary acclimatization. It would be necessary to adapt to the short period when food is abundant and to develop capacity to produce antifreeze substances (p. 37). The limited extent of shallow water, the prime habitat for fish diversification, would restrict opportunity for evolution of the requisite specialized physiology. Another group besides the notothenioides which has been successful in getting south of the Polar Front is that of the eel pouts. For these deep sea benthic fish the Polar Front is only a surface feature and the deep sea habitat is relatively uniform everywhere (Eastman 1993).

Another feature of Antarctic fish is that few species frequent the upper 200 m of the water column. The notothenioides do not possess swim bladders and about 50% of Antarctic species are found in benthic habitats. They are only secondarily pelagic. Some species (e.g. Antarctic cod, *Notothenia rossii*), one that has been extremely overfished in recent years, shifts its habitat during development. Larvae and fingerlings are pelagic, fingerlings and juveniles are benthic in beds of macroalgae, and adults are semipelagic in offshore feeding and spawning grounds. These semi-pelagic fish do not have adaptations for neutral buoyancy. Silverfish, an abundant ecologically important notothenioid with a circumantarctic distribution, is a truly pelagic species, shoaling, feeding, and spawning in open water down to 900 m. It achieves neutral buoyancy by reduction of skeletal material and accumulation of fat. Silverfish is the major fish species both in number and biomass in most shelf areas of the Southern Ocean. It feeds on copepods, euphausiids, and chaetognaths and is itself preyed on by most of the larger carnivores. The other abundant pelagic fishes in the Southern Ocean are the Myctophidae, lantern fishes, so-called because of their luminescence. They are small fish, migrating vertically from depths between 200 and 1000 m in the day to the surface at night. Having swim bladders, which are acoustically reflective, they are picked up by echo-sounders and are the cause of the 'deep scattering layers' which at one time greatly perplexed sonar experts. They are opportunistic feeders on crustacea and larval fishes. The biomass of these mesopelagic fish is said to exceed that of krill, benthic fish, birds, seals, and whales, and they are undoubtedly an important component of the food web (Knox 1994).

Surveying what is known of the ecology of Antarctic fishes one is left with the impression that the Southern Ocean is underutilized by fish and that possible niches for them are unfilled.

8.3.3 Seabirds

After a long period in which seabirds were observed for themselves alone they are now recognized as highly adapted and quantitatively important components of the pelagic ecosystem. This is true world-wide but much of the research leading to this new viewpoint has been done in the polar regions.

The diversity of birds frequenting the open waters of the Arctic is low, about 95% of those breeding in the Arctic belonging to four species – the northern fulmar (*Fulmarus glacialis*), the kittiwake (*Rissa tridactyla*), Brünnich's guillemot or thick-billed murre, and the little auk (Sage 1986; Pielou 1994). These breed on land in colonies which are often huge (see Fig. 4.21). Arrival at their breeding grounds is synchronized with the breakup of ice so that zooplankton and young fish are available when young are hatched. A colony of little auks containing 100 000 pairs transports some 71 tonnes of zooplankton from seas to colony during the four weeks of summer. Distribution of the birds at sea is related to local abundance of prey. Polynyas (p. 160), being areas of upwelling, are productive and particularly important, when they remain open, for overwintering. Fronts (p. 24) provide zones of high productivity and are marked by concentrations of seabirds, Lancaster Sound (*c.* 74°N 85°W) is an area where several million birds congregate in the summer. Of the eight species found there, three, the northern fulmar, kittiwake, and Brünnich's guillemot, feed largely on the Arctic cod. Brünnich's guillemot dives, wing-propelled, to depths of 50–75 m for two to three minutes overall after its prey. The cod is sustained by algae growing on the undersurface of the sea ice. This, then, is a simple food web with the flow of energy channelled mainly through one species of fish.

In the Antarctic, there are two main groups of seabirds, the procellariforms (albatrosses, petrels, etc.) and the penguins, which include the most highly adapted among birds to marine conditions (Croxall in Laws 1984; Knox 1994; Williams 1995). The wandering albatross (*Diomeda exulans*) navigates and survives in immense areas of ocean. It is a large bird, weighing up to 10 kg and with a wing span of up to 3.5 m, which spends most of its adult life at sea and may live to be 70 or 80 years old. Its flight is seemingly effortless, using updraughts to give it height to glide with minimum movement of its stiffly held, narrow, wings. It seems that it is only in latitudes between 40 and 50°S that albatrosses can find sufficiently strong and constant winds for this mode of flight. Dependence on wave-generated updraughts when flying near to the sea surface probably accounts for their absence from the pack ice (Fig. 8.5). The energy demands for seabirds in swimming, flapping flight, and gliding flight have been roughly estimated as, respectively, four, six, and two times resting metabolic rate. Observations on ringed birds show that they can accomplish journeys of over 5000 km in 10 days. Using satellite telemetry to track birds equipped with small transmitters, foraging trips from nests on Îles Crozet (45°S 52°W) have been followed with greater accuracy. During these trips, which may cover between 3600 and 15 000 km, they fly at up to 80 km h^{-1} over distances of up to 900 km day^{-1}. They remain active at night, this being an advantage since squid, their principal prey, are luminescent: prey is seized from the surface. As is obvious to the seafarer, these birds are also opportunistic scavengers. The excess salt that they inevitably take aboard while feeding is disposed of by nasal salt secreting glands. The flight path and foraging strategy are determined by the wind. When following the wind they fly at a slight angle to the left, which, in the southern hemisphere, leads them away from cyclonic lows and towards high pressure. Like most Antarctic birds and sea mammals, the wandering albatross has little fear of humans. This makes the fitting of telemetric devices comparatively easy. Another device, which the bird is induced to swallow, monitors stomach temperature every 30 seconds. Since food taken in is at sea temperature and lowers stomach temperature in proportion to its size, the frequency of feeding at sea and amount taken can be followed: 600 000 tonnes of squid are estimated to be taken by seabirds, mainly by albatrosses and king penguins, breeding on South Georgia and islands of the Scotia Arc.

Penguins are as proficient as navigators as are albatrosses but, as yet, we are ignorant of their methods. Their characteristic feeding technique is pursuit diving. Miniature depth recorders, logging the number of dives and their depth and duration, have been used to study several species. King penguins (*Aptenodytes patagonica*) dive to depths of 236–265 m after fish and squid. Chinstrap and Adélie penguins, both krill eaters, dive mostly to shallower depths, around 10 m, with 70 and 170 m maxima, respectively. The Adélie penguin catches krill at an average rate of 7.2 g per minute spent under water. Radiotelemetry has been used to monitor behaviour at sea. The maximum overall swimming speed is around 5.5 km h^{-1}. Foraging ranges can be estimated from nest relief intervals. Feeding times and travelling times are different for the various species; gentoo penguins (*Pygoscelis papua*) range within 17 km of the nesting site whereas chinstraps go further, up to 27 km. The gentoo requires more krill to rear its chicks than does the chinstrap, 118 as compared with 73 kg per breeding pair, and achieves this by having a more restricted foraging range and shorter nest relief intervals, as well as by diving more deeply. Such results explain how these birds, preying on the same organism and often sharing the same nesting sites, manage to coexist; their feeding niches do not overlap as much as might be supposed.

A tentative estimate of krill consumption by seabirds around South Georgia and the Scotia Arc is 10.9×10^6 tonnes per annum. There is no doubt that, quantitatively, they are an important element, comparable with other predators (p. 201) in the pelagic food web. Their predation must also play an important qualitative role in structuring pelagic communities. As has already been pointed out (p. 101) they have a profound effect locally on terrestrial habitats. Broadly, the ecological impact of seabirds seems similar in Arctic and Antarctic waters. An attempt to make quantitative comparisons around islands at roughly comparable latitudes north and south – in the Bering Sea in the area 55–63°N, 169–173°W, and the South Orkneys 61°S 45°W – bears this out, Numbers of birds per unit sea area tended to be higher in the Arctic but since Antarctic birds have a larger mean size the biomasses are similar. Because smaller birds require proportionately more food than larger ones, consumption of pelagic prey is probably greater in the Arctic than in the Antarctic.

8.3.4 Seals

Seals, being accomplished swimmers amply insulated with dense fur and blubber, are well adapted to polar waters. Nevertheless, they are visibly most associated with their breeding sites on land or ice. They are not easily studied in the open sea and precise information on their lives in the pelagic environment is sparse (King 1983).

Several of the Arctic species frequent shallow coastal waters and feed on benthos. The harp seal (*Phoca groenlandica*), by contrast, lives in the open seas of the Arctic Atlantic. It is a gregarious animal, migrating as far north as Cape Chelyuskin (77°44′N 103°55′E) in the summer and early autumn and south as far as Newfoundland in the later autumn. It breeds in the spring on the margins of large ice fields. While migrating, the harp seal jumps out of the water rather like a dolphin does. Pups feed on planktonic crustaceans and adults on shoaling fish such as capelin, herring, and polar cod. The total population is believed to be around 2.5 million – a considerable reduction from what it was before commercial exploitation. The hooded seal is not quite so extensively distributed in the same area. It is solitary at sea, feeding on Greenland halibut, capelin, cod, and squid, for which it may dive to about 180 m. The total population is perhaps rather less than half a million.

In the North Pacific, Chukchi, and Bering Seas, the predominant pelagic seal is the ribbon seal. After breeding on ice in the spring it disperses, scarcely anything being known of its movements thereafter. Fish and cephalopods appear to be its main food. The total population is now about a quarter of a million (Sage 1986; Ainley and DeMaster in Smith 1990).

In the Antarctic, Weddell, Ross, crabeater, and leopard seals are associated with ice and only the fur seal (*Arctocepthalus gazella*) and elephant seal (*Mirounga leonina*) are more characteristic of open water (Fig. 8.5). The elephant seal is circumpolar in distribution, although divided into three breeding stocks, and the fur seal probably was until it was exterminated on most peri-Antarctic islands. From near extinction in the 19th century the fur seal has made a spectacular recovery since around 1970, now totaling about a million animals and creating ecological nuisance on many islands. Krill, fish, and squid are its foods. Diving behaviour has been monitored by attached instrument packages which show a diurnal pattern with most diving taking place at night when krill are near the surface (Fig. 8.7). The depth of dives is mostly between 20 and 50 m with a maximum of 100 m. Little is known of what fur seals do between May and October, when they are absent from the breeding grounds. Individuals tagged on South Georgia have been seen on the South Orkneys, South Shetlands, and in South America on Tierra del Fuego. What sea elephants do between going to sea after moulting in March to April and returning to shore to breed in September has also been mysterious. The stomach contents of animals on breeding beaches are mainly squid and fish, prey similar to that of sperm whales, which feed at great depths. The retinas of elephant seals are rich in a visual

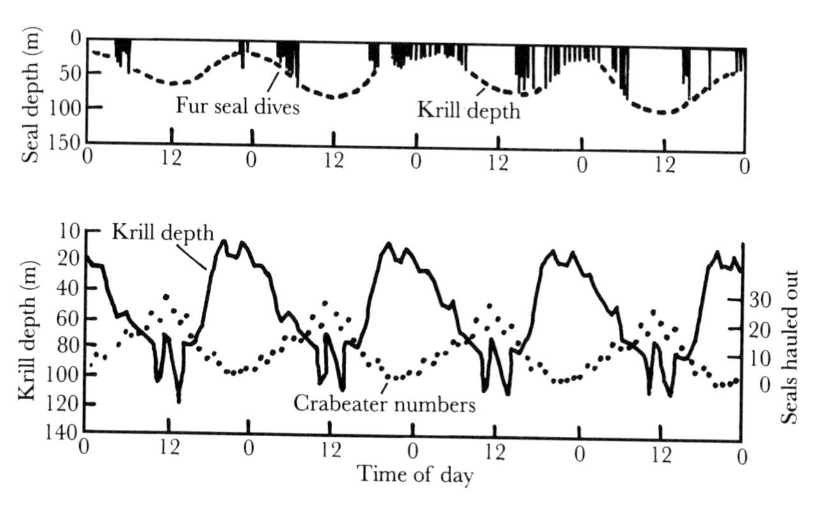

Fig. 8.7 Patterns of seal behaviour in relation to krill distribution; observations over four days on fur seal diving (above) and crabeater seal numbers hauled out on the pack ice (below). (Redrawn from Laws 1984, courtesy of the author and Academic Press, London.)

pigment similar to the 'deep-sea rhodopsin' of deep water fishes, suggesting adaptation for detecting luminescent squid. Successful fitting of elephant seals with time–depth recorders has begun to fill in some detail in this picture. At sea, they spend 80–90% of their time underwater, most of it at depths of 200–400 m but sometimes going down to 1500 m. It seems that they feed at the pycnocline, where particulate matter tends to accumulate and attracts squid. The time needed to take breath at the surface is astonishingly short and their muscles must be able to work without oxygen for much of the time when they are submerged. How they navigate, evidently with great precision, when travelling in deep water over the long distances between their breeding and feeding grounds, is not known. The food consumption of fur and elephant seals is estimated as 1.3 and 6.1 million tonnes per annum, respectively (Laws in Laws 1984; Knox 1994).

8.3.5 Whales

Here, at the apex of the trophic pyramid are individuals of around 100 tonnes fresh weight. Compared with those at the bottom, weighing less than 1 picogram, there is a difference of twenty orders of magnitude. To put things in perspective, the total biomass of bacteria in the photic zone of the Southern Ocean, assuming $1\,g\,m^{-2}$ (see Karl in Friedmann 1993) is about 3.2×10^7 tonnes, whereas the present biomass of whales is around 8×10^6 tonnes, only a quarter as much.

Whales, cetaceans, are of two kinds, the whalebone or baleen whales, Mysticetes, and the toothed whales, Odontocetes. Baleen whales feed by sieving out zooplankton, such as krill, by means of rows of hairy triangular plates carried on each side of the palate. They include the blue (*Balaenoptera musculus*), fin (*B. physalus*), minke (*B. acutorostrata*), and the right (*Balaena glacialis*) whales. The toothed whales, which feed on larger animals, include the sperm (*Physeter catodon*) and killer (*Orcinus orca*) whales. Most of these species are found in both Arctic and Antarctic seas but, because their life pattern is to feed in polar waters and breed in equatorial waters and there is therefore a six-month phase difference in breeding between the hemispheres, it seems that there can be little mixing of the stocks across the equator. In the Antarctic there is latitudinal zonation in the distribution of whale species (Fig. 8.5).

The baleen whales feed near the surface, rarely diving to any great depth, and probably finding their prey by both sight and echo-location. The technique of feeding varies; the right whale skims the sea surface, swimming slowly with jaws agape, filtering water as it goes. The blue, fin, and minke whales take gulps of water plus krill then, with mouth closed, the water is forced through the baleen plates by expansion and pushing forward of the tongue with contraction of the ventral grooves which run backwards from the chin. Whales may concentrate krill by encircling a patch then diving to come up vertically beneath it with open mouth, or, by swimming beneath the surface and releasing a trail of air bubbles in which the kill become collected. Whales caught off South Georgia had stomachs full, or nearly full, in 70% of cases as compared with less than 25% for those caught off South Africa. For a 14.7 m female sei whale (*Balaenoptera borealis*) 305 kg of krill constitutes a stomach-full. When feeding maximally, 3–4% of body weight is consumed daily.

Sperm whales feed largely on squid, which they evidently find by echo-location and it has been suggested that they can immobilize prey by a projected beam of sound. They can stay below the surface for as long as an hour. The maximum quantity of food found in their stomachs is about 200 kg but in one instance this consisted of one 12 m giant specimen of the squid *Architeuthis*. Again, daily consumption is about 3% of body weight. Little is known of the feeding of the smaller Odontocetes – killer whales, pilot whales, and dolphins. Killer whales chase their prey, using their sharp teeth for seizing, biting, and tearing. They hunt in packs and co-ordinate attacks on large prey such as seals, penguins, and even the large baleen whales. They do not dive deeply but take their victims at the surface (p. 170). Estimates of food consumption are 4% of body weight for killer whales, 4–6% for pilot whales, and 4–11% for dolphins. As with most animals, the relative food requirement increases as the size of the animal decreases.

Whales have a considerable impact on the pelagic ecosystems of polar regions. Before whaling reduced stocks so drastically, baleen whales in the Antarctic took an estimated 190 million tonnes of krill per annum and the sperm whale 10 million tonnes of squid, the corresponding recent figures being 43 and 4.6 million. Possibly because of a greater availability of food for those which have survived, baleen whales have recently shown increased growth rates, earlier maturity, and higher pregnancy rates than formerly. Population increases are not yet evident but stocks of fur seals and penguins have increased (p. 201), evidently in response to this stimulus. Because the large whales are migratory they export the biomass accumulated in polar waters to equatorial regions where they undergo almost total fast. This is a loss of some 18 million tonnes per annum of whale material from the Antarctic, and a corresponding enrichment in energy and nutrients of their breeding grounds (Brown and Lockyer in Laws 1984; Ainley and DeMaster in Smith 1990; Knox 1994; Pielou 1994).

8.4 The polar marine ecosystems

Pelagic ecosystems are not sharply defined. In the Arctic, the boundaries between polar and cold temperate waters are confused and although the Southern Ocean is demarcated by the definite line of the Polar Front, birds, mid-water fish, and whales cross this frontier without hindrance. There are exchanges between the pelagic and the sea ice, benthic, and terrestrial ecosystems. Dead organisms, faecal pellets, and other detritus falling from above are usually the main source of food for the benthos. Regeneration of mineral nutrients takes place on and in the sea bottom and they are returned to the water column above given suitable conditions. This intermeshing of pelagic and benthic ecosystems is locally important in the Antarctic but more extensive in the Arctic, where the areas of shallow shelf seas are greater. In coastal waters of the high Arctic the key pelagic species, Arctic cod and some of the larger predators, such as the ringed seal, depend equally on benthic and pelagic prey.

Polar seas have been commonly thought of as having the simplest of food chains – primary productivity (diatoms), to secondary producers (krill), to tertiary producers (whales). This is an oversimplification; primary producers other than diatoms are abundant and some of them are consumed by krill; krill can be carnivorous and at both these levels there are many other predators and alternative pathways for the flow of energy. As a basis for making predictions about standing stocks the concept of the food chain is usually useless. A food web, in which an attempt is made to include the ramifications of the trophic interrelations between all the important species in an ecosystem, is more realistic. Simple food webs for Arctic and Antarctic pelagic ecosystems can be cast in similar form (Figs 8.8a,b). The main groups of animals are mostly the same although the species are different. More elaborate schemes would include the ultraplankton and the microbial loop.

Both in the Arctic and Antarctic the primary productivity which provides the input of energy into the food web is low compared with other parts of the world's oceans (pp. 191 and 193). Primary productivity is greatest, and upper trophic level organisms are concentrated, in particular localities by shelf-break fronts, convergences, islands, marginal ice zones, and polynyas. In both polar regions the upper trophic levels include few species but these are represented by large numbers of individuals. The further concentration of these into restricted breeding sites gives an impression of high productivity that is misleading. There are about 4.5 mg of penguin per square metre of Southern Ocean – one-millionth of the concentration in breeding colonies. Estimates of ecological efficiency, the ratio of predator production to prey consumed, are difficult to make because of the paucity of data but tentative values are around 5%, less than the 10% which is usually assumed for temperate marine ecosystems. There is even greater difficulty in constructing mathematical models to describe energy flows as envisaged in Figs 8.8 and 8.9. A model of this sort for the Southern Ocean was one of the objectives of BIOMASS (p. 199), which occupied 12 research vessels for ten years, but the complexity of the behaviour and distribution of *Euphausia superba* and the difficulty of getting reliable, estimates of its biomass and activity have hindered progress with this. Models of parts of the food web can be useful, however. A simple model of population changes in krill and two of its predators, seals and whales, predicts the course of events under different harvesting regimes (May *et al.* 1979). This provides a useful approach to fisheries problems in general.

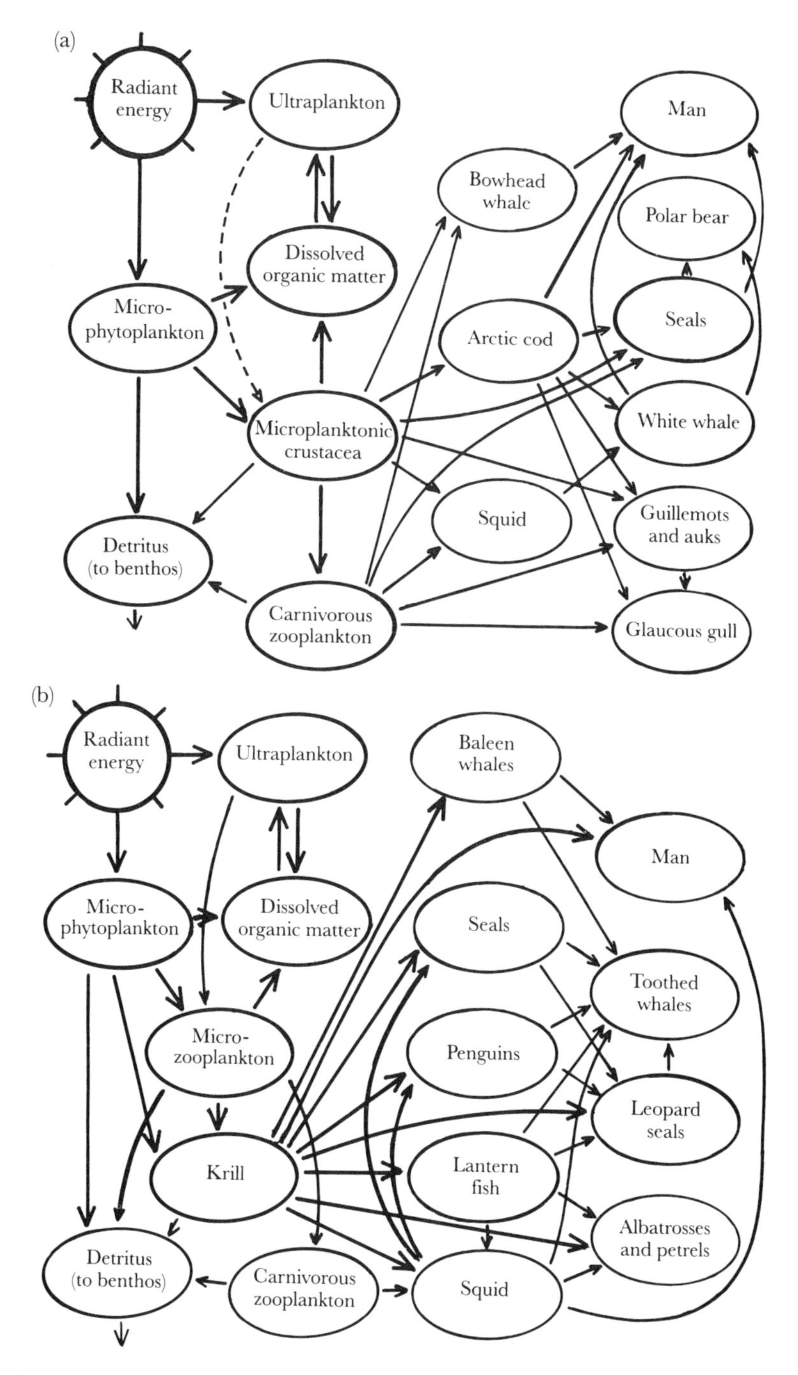

Fig. 8.8 Flows of energy and materials in the open seas of: (a) the Arctic; and (b) the Antarctic. The role of the ultraplankton in the Arctic is uncertain

Mineral cycling is coupled to the traffic in organic carbon and energy and corresponding webs can be drawn up for nutrient elements, with appropriate shifts of emphasis to different groups of organisms. The silica cycle is of particular interest since global mass balances indicate that in the Southern Ocean, dissolved silicate is converted to some 1.3×10^9 tonnes yr^{-1} of solid opal, resulting in over 75% of present-day biogenic silica deposition in marine sediments happening south of the Polar Front. This is to be contrasted with the 5% of the total marine deposition of carbon that occurs in the same area. Diatoms, which have heavily silicified cell walls, are particularly abundant in the Southern Ocean but in addition it seems likely that the low temperature of the surface waters, which reduces the solubility of silica, results in an unusually high fraction of the silica in dead diatoms sedimenting out (Jones *et al.* in Smith 1990). A cycle of profound importance for the global environment, is that of carbon. As a result of mankind's use of fossil fuels, the concentration of carbon dioxide in the atmosphere is steadily increasing and, probably, causing rise in global temperature (pp. 5 and 179). It seems that certain parts of the polar seas may counteract this by acting as carbon dioxide sinks. This can happen when highly absorptive cold water sinks and carries dissolved carbon dioxide to the bottom, or when particulate organic matter produced by photosynthesis sinks to be incorporated in sediments. Areas which may be important in both these ways are the Greenland Sea and the Weddell Sea. Any mathematical model designed to predict the course of global warming must take this possible biological effect into account but the necessary data are at present rather meagre (Tréguer in Hempel 1994; Priddle *et al.* in El-Sayed 1994; Gradinger in Wadhams *et al.* 1995). It is a measure of this inadequacy that a model of the Antarctic pelagic food system, which admittedly rests on some bold assumptions but is nevertheless plausible, suggests that the Southern Ocean, contrary to accepted opinion, is inefficient as a carbon sink. This inefficiency arises from the large flux of carbon returned by respiration, not to the sea, but to the atmosphere by air-breathing seabirds and marine mammals (Huntley *et al.* 1991).

The complex intermeshing of organisms and processes which constitute the ecosystem begins to be apparent but there are more intricacies to be taken into account if we are to progress towards even an approximate understanding. Organisms go through cycles of development and their activities, behaviour, and interrelations change according to the stage they have reached. These functions are geared to the periodicities of the physical and chemical factors in the environment and it seems certain that the term 'chemical' must include organic, biologically produced, substances as well as inorganic nutrients. The dissolved organic matter in seawater comprises a great variety of substances, many of which are biologically active. These have been variously called *ectocrines*, *ecomonies*, or *chemical telemediators* and are concerned in antagonisms, communalism, sexual activity, location of food, alarm signals, and other behavioural phenomena. For example, many marine predators are attracted by specific substances, often amines, released by their prey. Biochemical ecology is a developing subject that has scarcely touched the polar regions yet, nevertheless, there are clear indications that chemical mediation must occur in these waters. Fish in general can detect extremely low concentrations of dissolved substances and the olfactory apparatus which does this is well developed in Antarctic notothenioides although no experimental data on its possible role in feeding, migration, or reproduction seems to be available. Sonic vibrations provide another means of exploring the environment and communication between individuals. The oceans are loud with clicks, squeaks, warbles, whistles, moans, and bellows produced by organisms ranging from zooplankton to whales. Movements of the appendages of krill as they swim in swarms produce a vibrational disturbance which presumably is a useful guide to predators when vision is restricted. The lateral line, a short-range mechanosensory system in fish, responds to water currents and vibrations It is well developed in notothenioides, in which it has been investigated experimentally (Eastman 1993). Vibrations of a frequency around 40 Hz, produced by swimming crustaceans, stimulate the lateral line system of *Pagothenia borchgrevinki*. This is of value in detecting prey within 30 mm of the fish. The 'songs' of seals, dolphins, and whales, which attract so much popular interest, are means of communication which, if of a frequency less than 1 kHz, can carry long distances, perhaps as much as 16 km horizontally. Rapid sharp clickings in the range 5–32 kHz are additionally a means of echo-location. Such signals no doubt contribute materially to the workings of the ecosystem and man-made underwater noise must be regarded as a potentially damaging form of pollution. Tankers passing a vocal species closer than 100 m could impair or black out communications between animals more than 100 m apart (Stirling and Calvert 1983).

9 Mankind in polar habitats

9.1 The first invasions by man

Only recently have the enormous areas around the poles been invaded by mankind (Sugden 1982; Ives and Sugden 1995). The Arctic, being nearer to centres of population and approachable by land, was the first to suffer intrusion. This came via the unglaciated lowland of northeast Siberia and the Beringia peninsula (p. 47) which was occupied by groups of people hunting mammoth, bison, and caribou with stone-tipped spears and living in winter in semi-subterranean dwellings. This Palaeo-Arctic culture persisted until around 7000 BP but long before, perhaps 20 000 BP, it had spread across the then existing land bridge to North America. About 11 000 BP the climate became warmer, ice caps melted, and the land connection with Asia was cut by the rising sea, leaving the primitive Arctic hunters to evolve in isolation. Those that remained in coastal areas lived on sea mammals, muskoxen, and caribou, and gave rise to the palaeo-Eskimo who reached northern Greenland about 3000 years BP. Another group, the Dorset culture, with the same origin, moved to the eastern Canadian Arctic and developed the building of ice houses (igloos), boats (umiaks and kayaks), and efficient harpoon technology for hunting on ice. This exploitation of sea mammals was the key to successful occupation of the high Arctic. Another wave of migration from Alaska occurred around AD 1000 establishing the Thule culture, which extended as far as western Greenland and adapted Dorset techniques to hunt the bowhead whale (*Balaena mysticetus*). It was around this time that the Thule peoples came into contact with Europeans when Norsemen discovered south Greenland and established colonies on its west coast. The Thule culture, which was more advanced than that of their descendants, present-day Eskimo (Inuits), ended during the 16th century, perhaps through a combination of climatic deterioration, disease, and European whaling activities. Isolation, with no exposure to the pathogens rife in Europe meant that these people had no resistance to some common diseases – in the early years of the 20th century measles wiped out many local Inuit populations. The present-day Inuit are a homogeneous people speaking variants of the same language and clearly demarcated from the forest Indians of North America. Their way of life seems to be a recent makeshift adaptation to changing conditions. The western Inuit in the southern Bering Strait area were not affected by the Thule culture but nevertheless retain affinities with the eastern Inuit (Sugden 1982; Fitzhugh and Kaplan 1982; McGhee in Ives and Sugden 1995).

In the Eurasian Arctic populations have been more diverse in origin and have colonized sub-Arctic, rather than Arctic, habitats. The great rivers of Siberia allowed people, ideas, and technologies to travel northwards so that the pattern and complexity of northern Eurasian communities stands in contrast to those of the isolated peoples of North America. The principal mode of life which evolved was a migratory one, following the reindeer herds which provided their sustenance (p. 222).

The Antarctic, isolated by the broad tempestuous barrier of the Southern Ocean, was invaded by man only in the last two hundred years (Fogg in Ives and Sugden 1995). The first recorded landing south of the Antarctic Polar Front was when James Cook and his shipmates set foot on South Georgia on 17 January 1775. On the continent itself the first landing was probably in the Hughes Bay area of the Peninsula from the US sealing tender *Cecilia*, captained by John Davis, in February 1821. It may be that Polynesian seafarers penetrated into Antarctic waters before this – a Rarotongan legend recounts a voyage into a region of fogs, monstrous seas, and what may have been icebergs and brash ice, but they evidently did not establish settlements. Two projectile heads of a type used until about 1500 by the Indians of central Chile have been dredged from Admiralty Bay, King George Island (62°10′S 58°30′W). The natives of Tierra del Fuego, of the same ancestry as the ancient colonizers of the Arctic, were certainly hardy, accustomed to going naked in frigid conditions, and might have reached this spot in their canoes. However, the manner of arrival of these artefacts in Antarctica has its dubious aspects.

9.2 The ecology of pre-industrial man in the Arctic

In the 1950s, the total population of Inuit in Greenland, Arctic Canada, and Alaska was around 50 000. This may be taken as an upper limit to past populations and considering the mobility which was the essence of their culture, it seems that their impact on ecosystems must have been light. Their small communities were self-sufficient and depended only on renewable resources. If food became scarce locally then the Inuit moved elsewhere. Mobility required a minimum of possessions and Inuit equipment, notably the kyak, constructed with a frame of bones or driftwood joined by walrus hide thonging and covered with seal skin, is a remarkable combination of lightness with robustness. Sledges could, if necessary, be contrived from frozen fish. Living in sub-zero temperatures was made possible by the anorak and parka, the efficiency of which is attested by their present universal adoption as suitable wear for activities involving any suggestion of exposure to bad weather. The winter habitation of the northern Canadian Inuit, the igloo, constructed of blocks of firn, is pleasant and warm to live in. It has been found by the Swedish army to be proof against mortar fire and invisible to infrared sensors. Unfortunately, the early explorers from more advanced parts of the world learnt little from Inuit techniques for survival.

The ancient hunters of reindeer in the Eurasian Arctic perhaps numbered no more than 11 000 if it is accepted that the number of wild reindeer in the tundra was about 3 million with an annual increment of 7%. Exploitation of the reindeer involved a nomadic existence, following them south in winter from open tundra to forest tundra and northern taiga, where firewood was also to be found. Reindeer numbers imposed a limit on the human population and migrations distributed the impact on the main winter fodder, lichen, allowing its regeneration so that ecological balance was struck. With development of reindeer husbandry people were able to settle in the sub-Arctic, hunting of other animals and fishing began and populations increased. A rough estimate of the number of people is 30 000 in the Russian far north practising the indigenous way of life before the Russian conquests. This population had no large-scale or irreversible effects on the vegetation cover but minor changes in flora and fauna were brought about (p. 223).

The west coast of Greenland was settled by Norsemen in AD 986 and at the height of its prosperity the colony contained about 3000 in 280 farms. The main livelihoods were trade in walrus skins and ivory, and cattle-rearing. With deterioration in climate and, possibly, introduction of diseases from Europe in the 15th century, the colony became extinct. The ecological impact of this colonization was localized but some 50 vascular plants in the present flora were probably introduced by the Norsemen. Recolonization by Denmark was begun in 1721 with a policy of establishing contacts with the outside world but reserving all natural resources for the Greenlanders themselves. In 1805 the population was 6046 individuals.

9.3 Sealing and whaling

The taking of seals and small cetaceans for food has been practised since neolithic times but commercial whale fishery involving long-distance voyages to the Arctic did not develop until the 17th century. Willem Barents, who discovered Svalbard in 1596, reported abundant whales in its vicinity and this attracted the attention of Basque fishermen who had for a long time been hunting the Atlantic right whale in European waters. Soon, English whalers, followed by Dutch and Danish, were operating in the Svalbard area. The whales were pursued in small boats and harpooned by hand. Much of the whaling took place along the ice margin, which shifts from day to day and varies greatly in position from season to season. Environmental conditions required expert interpretation and it is significant that William Scoresby Junior, a scientist whose book on the Arctic regions (1820) was to become the foundation of Arctic science, was also the most successful whaling captain of his time. Initially, whale oil, in demand for illumination, was the only product that was valued and carcasses were discarded once the blubber had been removed. The two whales fished in this period were the bowhead, which frequents the ice fields, and the right whale, which avoids the ice. Whales became scarce around Svalbard by the early 18th century and whaling shifted to the Davis Strait off western Greenland. There, too, stocks showed a gradual decline. Statistics of Dutch whaling illustrate this. In the decade from 1661, when records began, the number of whales caught totalled between 452 and 1013 per year and between 3.4 and 7.3 per ship. In the decade 1761–70, after which the size of the fleet declined because of intensified competition, the corresponding figures were 179 and 1131 whales per year, with between 1.1 and 7.4 caught per ship. There was a similar decline in the Bering Strait fishery from an estimated stock of bowhead whales of between 9000 and 18 000 in 1848 to about 2300 in 1978. Pressure on the whale populations intensified with the development of the harpoon gun in the 1860s and the introduction of steam-powered whale catchers towards the end of the century. These enabled the large blue and fin whales, previously beyond the scope of the hand-propelled harpoon, to be taken but nevertheless the northern whaling grounds became unprofitable and attention was turned to Antarctic waters.

Seals, including particularly the walrus, were taken for blubber and ivory in the early years of commercial whaling in the North Atlantic. However, whales were the more valuable and it was not until whales became scarce that seals were again included in the catch. At the same time harp, hooded, and bearded seals were being taken in quantity for their skins by professional sealers.

The northern fur seal (*Callorhinus ursinus*) was discovered by Pribilof in 1786–7. At that time the stock probably numbered around 2.5 million. The fur being highly prized, especially in the Chinese market, the seals were soon being slaughtered, first by Russian and then by American sealers. In the 20 years up to 1909, 600 000 skins were taken and as many more seals lost after being wounded, so that the total population was reduced to about 300 000. Other species of fur seal in temperate waters suffered similar onslaught and by the beginning of the 19th century the South American fur seal (*Arctocephalus australis*) was becoming scarce. Sealers quickly moved in when fresh sealing grounds were reported or rumoured and following Cook's landing on South Georgia, British and American vessels soon frequented its waters. Fur seals were the main quarry but elephant seals, which yielded oil of a similar quality to that of whales, were also taken. Sealing operations were ruthless and, although some of the more thoughtful sealers realized that their methods were self-defeating, without the slightest regard for conservation of stocks. Following the discovery of the South Shetland Islands around 1820 there was frantic slaughter of their fur seals but by 1822, when nearly 1.25 million of them had been killed, the species was commercially extinct. Through the 19th century sealing, mainly for elephant seals, continued intermittently as stocks recovered, but never reached the intensity of the 1820s. By the end of the century it seemed that the Antarctic fur seal was biologically extinct. However, a few did survive and following occasional sightings in the early years of the new century there was a population explosion, beginning on the South Georgian islands around 1970 but quickly spreading to the South Orkneys and South Shetlands. The current total population is probably over a million. This spectacular recovery may be put down to freedom from hunting and increased availability of krill. The elephant seal was not persecuted so severely since the taking of small numbers for oil was not profitable but, it too, has increased in numbers in the second half of this century (Laws in Laws 1984; Knox 1994).

James Clark Ross on his voyage with the *Erebus* and *Terror* in 1839–43 reported right whales as abundant in the Southern Ocean. None were found by reconnoitring whalers in the 1890s. It seems unlikely that Ross, who was familiar with the right whale in the Arctic, was mistaken and it may be that hunting of this species in temperate southern waters had reduced the stock to a low level by 1890. The whales that were abundant, blue and fin, were vulnerable to the new techniques. Commercial whaling began in earnest with the establishment of a whaling station at Grytviken on South Georgia in 1904. This flourished and was quickly followed by others in the South Atlantic sector. The number of whales of all species caught from South Georgian stations rose to over 7000 per year by 1916. Recognizing that this area of the South Atlantic was probably the most profitable whaling ground in the world and that, like the Arctic, it might quickly be fished out, the British government constituted it as the Falkland Islands Dependencies in 1908 and took steps to regulate the industry. Being able to control factories on land it was possible to limit their activities and to prohibit the taking of females with calves. Research on the factors controlling whale abundance, financed by a tax on oil production, was carried out by *Discovery Investigations* (p. 199). These measures might have saved Antarctic whale stocks had it not been for the development of factory ships capable of processing the carcasses at sea outside the jurisdiction of government. The first of these ships began operating in 1925 and by 1930 there were 41 of them. Factories on land progressively closed down. Between 1925 and 1931 the number of whales killed per year rose from 14 219 to 40 207. An international Convention for the Regulation of Whaling came into effect in 1935. Germany and Japan refused to adhere to the Convention and only Norway and the United Kingdom among the pelagic whaling nations observed it. In 1938, a whale sanctuary was designated in the area south of 40°S between longitudes 70°W and 160°W and complete protection for the humpback whale (*Megaptera novaeangliae*) was agreed. The International Whaling Commission has scarcely realized the hopes it raised when it was set up in 1946. Accurate quantitative data on which to base decisions was lacking for many years and the Commission could not enforce its decisions. A recent (1994) agreement established an extended Antarctic whale sanctuary with northern limits at

40°S in the east Atlantic and west Pacific and at 60°S elsewhere with the whole of the Indian Ocean as a breeding sanctuary. It remains to be seen whether the most hard-pressed whale species have been saved.

Following the establishment of the stations on South Georgia, catches of humpbacked whale were the first to rise and then fall, thereafter fluctuating at a low level with the prohibition on hunting it having no perceptible effect. There was little respite during the First World War and the 1930s showed catches of blue and then of fin whales rising to high levels followed by steep declines as the stocks were depleted. The Second World War allowed some recovery and then history repeated itself, the blue whale fishery being the first to collapse, followed by the fin whale in the 1960s. This left the smaller sei whale as the main quarry but this too was soon fished out leaving the minke whale, with a weight of 19 tonnes at the most as compared with the 150 tonnes of the blue whale, to sustain whaling operations. The present state of the whale stocks and the prospects for recovery are obscure – the decline of whaling has meant that accurate data on numbers are less available. An estimate of blue whale numbers in 1989 was 453 but recent figures have been higher, raising the hope that it may have escaped extinction. The most abundant whale in Antarctic waters is now the minke, with about 760 000 individuals (Brown and Lockyer in Laws 1984; Knox 1994).

In recent years, strong feeling against hunting seals has developed. This is usually based on a confusion of the desirability of maintaining population levels on the one hand and the inhumanity of hunting on the other, with little regard for the ecological background. The first international agreement for controlling exploitation was the North Pacific Fur Seal Convention of 1911. Hunting of other seals has been regulated around the coasts of Britain, Norway, Sweden, and Russia in the last few decades. A collapse of the world market for seal skins has discouraged commercial sealing but there is strong support, particularly in Canada, for aboriginal hunting to be allowed. For the Antarctic community, a pilot sealing expedition from Norway in 1964 rang alarm bells and led to a Convention for the Conservation of Antarctic Seals being ratified in 1972. The disappearance of the market for seal products rendered this Convention unnecessary but it was a valuable precursor of much wider measures for the conservation of Antarctic marine resources (p. 236).

9.4 Hunting

The first human colonizers of the Arctic were hunters. Although they were few in number and operated in an enormous area, their impact on the numbers of some animal species was probably drastic. Polar animals may exist in large herds but this is the result of low rates of annual increase taking place in a relatively stable ecosystem. The appearance of a new predator may quickly eradicate a species which has the habit of congregating in large numbers, as dramatically illustrated by the case of the Antarctic fur seal (p. 217). On the American continent, fossils show no decline in diversity or territorial range of large mammals until the spread of human invaders began around 11 000 BP. The extinctions happened suddenly. Similarly, some of the ancient animal inhabitants of the Eurasian tundra seem to have been exterminated by small numbers of hunters. Some palaeolithic sites in northern Eurasia contain astonishing quantities of the remains of slaughtered animals. One at Mezhirich in the Ukraine yielded bones representing 95 individual mammoths (*Mammuthis primigenius*). More recently, the great auk has been hunted to extinction and the walrus and muskox seriously threatened.

Whereas the aboriginal peoples took animals, more or less, at subsistence level the expansion of European exploration led to hunting on a much larger scale. The centuries-old Russian fur trade became a state-supported monopoly and by the end of the 17th century operated throughout nearly all northern Siberia. The Hudson's Bay Company, granted a charter in 1670, traded at first marginally in the Northwestern Territories, then soon reduced the supply of furs by unregulated slaughter. Its activities, together with those of other hunters from the south, extended into Arctic regions in the 18th and 19th centuries. Steel knives, guns, and patent traps enabled the Inuit to supply furs more easily but, of course, reduced the stocks. There was rapid rejuvenation of the North American Arctic fur trade after the First World War but no species was hunted to extinction. Public sentiment has now turned against the wearing of furs and the most threatened animals have been protected by legislation and the establishment of reserves (McGhee in Ives and Sugden 1995).

Apart from the hunting of fur and elephant seals (p. 217), there has been no important activity in the Antarctic which needs to be considered here.

9.5 Fishing

The indigenous peoples of the Arctic have always fished at the subsistence level. Impelled by increasing demand and exhaustion of local stocks and aided by retreating ice and long-distance fishing vessels, commercial fisheries in the North Atlantic and North Pacific have extended as far as they can into Arctic waters. A major expansion took place in the 1950s, the Barents Sea and the coasts of Iceland and Greenland being intensively fished, particularly for Atlantic cod. Competition led to the 'Cod War' between Britain and Iceland in the 1960s and 1970s. Iceland adopted the Exclusive Economic Zone principle, enabling its own fleet to catch more. Various conservation measures have been agreed by the nations fishing in the Arctic, including quotas, net mesh sizes, and restricted areas. Nevertheless, overfishing has happened. Catches in the North Atlantic reached a plateau of around 3.5 million tonnes in 1974 and both herring and cod have formed declining proportions of the total. Possibly this is a factor contributing to a decline in seal populations which has been noted in recent years. There are major fisheries for halibut and Alaska pollock in the Bering Sea, again showing considerable post-Second World War growth. Japan and the former USSR/Russia have dominated this fishery. Here, too, there have been impacts on populations of the top predators. In Norton Sound, Alaska (64°N 163°31′E) numbers of guillemot (common murre) began to decline in 1975, showing a distinct relationship to a decrease in catch per unit effort of the pollock fishery in the area where the birds spend the winter. There has been some development of fisheries around the Arctic Ocean itself. Herring, capelin, Arctic cod, and various flatfish have been taken to some extent but the main commercial fisheries in the Alaskan and Russian sectors have been based on salmon, whitefish, and Arctic char, anadromous fish using rivers for spawning. The indications are, however, that fish production in the Arctic ocean is low because of slow growth at near-zero temperatures and heavy predation by birds and marine mammals (Zenkevitch 1963; Ainley and DeMaster in Smith 1990).

The same pressures that forced fishermen north sent them south, although, the distances being so much greater, government subsidies have been needed. Sealers, whalers, and explorers caught Antarctic fish for the pan but commercial exploitation did not begin until the late 1960s when the imposition of 200-nautical mile (370 km) fishing zones elsewhere caused the redeployment of distant-water fleets, mostly Soviet at first. The total annual catch around South Georgia rose rapidly from a few hundred tonnes in the 1967–68 season to a peak of 400 000 tonnes in 1969–70. Catches were mostly *Notothenia rossii*. Fishing extended to the vicinity of other islands – Kerguelen, South Orkneys, and South Shetlands – and to other fish – notably the icefish *Champsocephalus gunnari* – with other nations joining in. Catches faltered but rose with increasing effort. However, each rise in catch in a particular locality or of a particular species was followed by a decline. By 1990, the stock of the most exploited species, *N. rossii*, in the Atlantic sector was less than 5% of what it was when fishing started. The Convention on the Conservation of Antarctic Marine Living Resources (CCAMLR) was concluded in 1980 but stringent control measures have been applied only since 1989. There is some evidence that stocks are now beginning to recover (Knox 1994).

Krill (p. 197), large enough to be caught with trawls and rich in protein, was another obvious resource to be harvested. The former USSR began looking into this in 1962 and by 1970 began marketing krill-based foods. The Japanese followed in 1972 whereas in Britain, which had done most of the research on krill thus far, activity was limited to speculations in the popular press on krill and chips as the food of the future. Soon, catch rates of 40 tonnes per hour were possible and the total catch approached 600 000 tonnes per annum. However, difficulties emerged. Because autolytic enzymes break down krill proteins rapidly, a catch intended for human consumption must be processed within three hours. Furthermore, the exoskeleton contains toxic amounts of fluoride. For these reasons and the cost of working at so great a distance from home, the krill industry has not expanded and there seems no immediate threat to stocks. Obviously, unlimited harvesting of krill might well be disastrous ecologically. Its conservation is covered by CCAMLR but the knowledge of krill population dynamics necessary to implement protective measures is inadequate. One of the main objects of the BIOMASS programme (p. 199) was to remedy this but, while much was learnt, it is still not possible to identify crucial points for conservation practices (Miller and Hampton 1989; Knox 1994).

Fishing often damages other components of the marine ecosystem besides species of commercial value. Recently, a fall in numbers of wandering albatross has been traced to the mortality caused when the birds take the bait on the long lines used for squid fishing and are dragged below the surface. One boat can kill as many as 200 albatrosses in a few days fishing and the breeding rate of the species is insufficient to keep pace with this.

9.6 Pastoral and agricultural development

The pastoral way of life in the sub-Arctic centres on the reindeer (Chernov 1985; McGhee in Ives and Sugden 1995). In Eurasia hunting of the wild animals changed into husbandry. Until the middle of the 20th century the Lapps followed traditional patterns; during the summer entire families lived with their herds in the tundra and retreated with them to forest tundra for the winter. Now reindeer tending is motorized using skidoos, motorcycles, jeeps, and helicopters operated from fixed bases. In the Russian Arctic, where herding was collectivized under the Soviets, domesticated reindeer number about 2.5 million as against 600 000 wild animals. The latter occupy terrain on which it is not profitable to graze domestic reindeer, which cannot survive on extensive but poorly productive pasture without causing permanent damage. In the North American Arctic ground reindeer (caribou) were hunted by the Inuit in summer and Indians in winter. Numbers were drastically reduced when firearms became available, falling from around 1 750 000 at the beginning of the century to around 200 000 by 1950. Some Inuit groups starved. An effort was made to reintroduce reindeer in the Northwest Territories but it was a slow process. Reindeer herding was seen, both there and in Alaska, as a suitable occupation for the Inuit and commercial outlets were organized. However, in the 1940s and 1950s reindeer numbers crashed because it had not been realized that lichen tundra can carry only one or two to the square mile (2.6 km^2). The herds ran out of control of the low man-power available and seriously damaged their habitat. By the 1980s the populations of free-ranging reindeer in both North America and the Russian Arctic were tending to increase. There do not now appear to be any adverse effects of herding on the tundra vegetation. Indeed, by providing a focus for conservationist concerns as both a ‘wild’ animal and the foundation of a subsistence economy in both Alaska and northern Canada, the reindeer may be helping in the preservation of the tundra in general (p. 226; Paine 1988; Vitebsky 1989).

Effects of agriculture on polar habitats are small. Agriculture is impracticable in the Antarctic and in the Arctic is restricted to the sub-polar fringe. A grazing economy existed in west Greenland during the 11th to 15th centuries and today is possible together with the raising of crops for local consumption in favoured areas of Alaska and Siberia. Efficient horticulture can only be carried out under glass with heavy consumption of energy and is not likely to be extensive enough to have appreciable effects.

9.7 Introduction of alien organisms by man

Inevitably, human colonizers and visitors to the polar regions take with them, deliberately or inadvertently, a variety of microorganisms, plants, and animals. Unfilled niches have allowed many of these to establish themselves, sometimes without much disturbance of pre-existing communities but sometimes with devastating effect on ecological balance. Nevertheless, man may often be doing no more than accelerating invasions which would occur anyway.

9.7.1 The Arctic

Introductions into the Arctic have been more extensive and have occurred over a much longer time-span than have those in the Antarctic. However, for both geographic and ethnological reasons they are unevenly distributed around the Pole. In North America and Greenland the Inuit, hunters travelling light, took a minimum of living plant and animal material in their baggage and one can only guess what microorganisms may have accompanied them. Their sledge dogs, domesticated wolves, do not revert to the wild in the high Arctic. The people going to the North Atlantic islands were European and some of the affinities of floras and faunas with those of Europe may be accounted for by accidental transport with animals, merchandise or ship's ballast. In Greenland, a group of species dependent on European farming practice established themselves around the Norse farms, mostly to become extinct with the demise of the settlements, but some surviving to become part of the west Greenland flora or fauna. This continues at the present day but human responsibility must not be too readily assumed – the presence on Arctic islands of some flightless insects of European origin has been shown to predate the arrival of man.

The Eurasian Arctic peoples, having more ready exchange with those further south, have brought about more introductions. For centuries the Chukchi of northeastern Siberia used a variety of plants in their diet and dunghills around their campsites support many species otherwise absent from the locality. In areas recently brought into cultivation one finds some species unintentionally transported mixed with seeds or food products. At

Noril'sk (69°21′N 88°2′E), following ploughing of hayfields 19 species of plant appeared, some hundreds to thousands of kilometres from their native habitats. Such introduced species are usually found only in the vegetative state and few acclimatize and persist in Arctic habitats, the acidic conditions in the tundra being inimical. However, some, for example, pineapple weed (*Matricaria matricarioides*) and groundsel (*Senecio vulgaris*), spread quickly into wasteground in human settlements. Others are able to establish themselves in intrazonal habitats such as shore lines, sand bars, and among boulders, where competition from native species is minimal.

Among animal introductions are flies associated with man. Some, for example, the housefly (*Musca domestica*), are found only in human company but others, such as the carrion flies, *Protophormia terranovae* and *Cynomyia mortuorum*, are widely distributed in the sub-Arctic and extend into the Arctic where human refuse is available. Few of these species propagate to any important extent in the Arctic. Among birds, the commoner species frequenting settlements, for example, the white wagtail (*Motacilla alba*), and wheatear (*Oenanthe oenanthe*), are widely distributed native species and not introductions. Sparrows, swallows, and starlings following man into northern latitudes are limited to settlements in the southern sub-Arctic. Rats and mice do not seem to have established themselves to any noticeable extent in the Arctic (Chernov 1985).

9.7.2 The Antarctic

Soon after their discovery, the peri-Antarctic islands were colonized by plants and animals brought by sealers and other seafarers (Leader-Williams 1985). A measure of the outcome is given by the number of alien flowering plant species in the South Georgian flora, which is 25 as against 17 native species. Rats, rabbits, cats, pigs, sheep, cattle, and reindeer have been introduced on various islands, only Heard and Macdonald Islands among the sub-Antarctic groups remaining free of alien mammals. These introductions can have profound effects on indigenous floras and faunas. In 1904 the Kerguelen cabbage (*Pringlea antiscorbutica*) was reported as virtually eliminated from the main island of Kerguelen by domestic rabbits left deliberately by HMS *Volage* during a visit to observe the transit of Venus in 1874. The Kerguelen rabbit (*Oryctolagus cuniculus*) is now distinct from its ancestral type but has not reverted to the original wild phenotype. Most introductions of domestic animals have failed, probably because of the rigorous climate. Some inadvertent introductions have succeeded only too well. Rats and mice have affected both plants and birds on South Georgia (p. 98). Efforts to eradicate rats, rabbits, and feral cats from individual islands have been unsuccessful and, if they had succeeded, it is doubtful whether the floras and faunas would have gone back to their original states.

The introduction of reindeer on to South Georgia by whalers in 1911 and again in 1926 established flourishing herds which have been subjects for intensive study (pp. 31 and 98). Having been contained within a limited area by glaciers the ecological damage they have done is small considering the island as a whole but with global warming and glacial retreat this begins to change. The successful introduction of brown and brook trout into Kerguelen freshwaters seems to have had no drastic ecological repercussions. Were fish farming to develop on Kerguelen the impact would be mostly on the inshore marine system, which would presumably be used as a source of food for the fish.

Introductions on the continent itself have been negligible. The alien grass, *Poa pratensis*, can germinate and grow under the adverse conditions of Cape Hallett (72°19′S 170°16′E) but has not established itself there. Sledge dogs were present on the continent continuously from 1945 until 1995. Although causing mayhem in penguin rookeries if let loose they have never become feral or a threat to Antarctic ecosystems. The Agreed Measures for Conservation of Antarctic Flora and Fauna (1964) under the Antarctic Treaty proscribe the deliberate introduction of species not native to the Treaty area but the enforced removal of sledge dogs from the continent under this ruling can only be regarded as a deplorable example of bureaucratic correctness. The conservation of microbial ecosystems is a different matter. Man inevitably brings with him an assortment of microorganisms and many of these remain dormant under Antarctic conditions to resume activity if conditions become favourable (p. 54). It is impracticable to prevent this and the Agreed Measures do not attempt to do so. Improvements in waste disposal at Antarctic stations reduce contamination somewhat but special measures have to be adopted in areas of particular microbiological interest. Some 20 Specially Protected Areas have been designated in various places in Antarctica, entry into which is only by special permit and surface vehicles are prohibited. Sites of Special Scientific Interest (SSSIs), set aside for research, are not so rigorously protected but one of them, Barwick Valley (*c.* 77°S 161°E), an example of a polar desert ecosystem, has been kept virtually undisturbed with special care to avoid introduction of non-native microorganisms.

9.8 Mineral and oil extraction

Most polar explorers hoped to discover valuable mineral deposits. Even as lately as the 1950s a general assumption amongst scientists was that the polar regions are there to be 'developed' (i.e. exploited for geological resources regardless of other considerations). This has changed.

The earliest profitable mine in the Arctic was that for cryolite at Ivittuut in Greenland (61°15′N 48°15′W), which had – it is now worked out – the only extensive deposit of this mineral. Cryolite is used in the electrolytic production of aluminium. Coal deposits on Spitsbergen were first mentioned in 1610 but not mined on a large scale until the beginning of the 20th century. Mining by Norwegians and Russians goes on at places near the sea and environmental damage is localized. If mining continues, new developments will be inland, leading to damage to areas at present untouched. The necessary roads will inevitably open up more of the country to tourism.

Since the 1960s, significant reserves of oil, natural gas, lignite, uranium, iron, lead, zinc, copper, nickel, manganese, precious stones, and gold have been identified in the Arctic (Miles and Wright 1978; Walton in Ives and Sugden 1995). Extraction presents great difficulties in logistics, finance, and politics. Nevertheless, it has taken place, for example the zinc-lead mines opened up on Little Cornwallis Island (p. 115) as an assertion of Canadian sovereignty, causing unforseen environmental damage. The exploitation of deposits in Siberia made the former USSR self-sufficient in minerals and could therefore be justified on political grounds even though it was uneconomic in terms of world economy. The Kola Peninsula (*c.* 68°N 34°E) is the most heavily industrialized area. Nickel, aluminium, iron, copper, zinc, and manganese are mined and refined with extensive contamination by waste heaps and toxic fumes.

The extensive oilfield at Prudhoe Bay (*c.* 70°10′N 148°30′W) on the Arctic Slope of Alaska was discovered in 1968 and began production in less than a decade. Materials for construction of the rigs and other installations were brought in summer by sea via the Bering Strait and in winter by the biggest air-lift in the history of civilian aviation. The most economic method of transporting the oil proved to be by a 1289-km long north–south pipeline, traversing tundra, forest, glaciated mountains, and rivers, to Valdez on the coast of Prince William Sound. At a cost of US$7.7 billion for pipeline and terminal it was the most expensive civil engineering project ever attempted. Among the many problems were those arising from permafrost, which in the Prudhoe area is some 600 m thick. First, it was necessary to determine corrections for its effect on the seismic reflection surveys of the oil-bearing deposits. Because permafrost prevents drainage, installations had to be sited, if possible, on better-drained islands in the tundra. The oil having to be kept warm during its passage, the pipeline had to be refrigerated where it was buried in unstable permafrost soils, otherwise it was carried above the permafrost on supports. It was a condition that the tundra should not be disturbed during construction and this was achieved – after a fashion! – by building gravel roads and pads about 1.5 m thick on top of it. The pipeline crossed the migration routes of reindeer and at the insistence of environmentalists it was raised in appropriate places sufficiently for the animals to pass below. The reindeer did at first make use of these underpasses but wolves were quick to size up the situation and the herds, while tending to avoid buried sections, now pass beneath the line anywhere except at the points arranged for them (Sheldon 1988).

The exploitation of offshore oil and gas fields in the Arctic is equally difficult and expensive. Ice is sufficiently thick in winter to allow heavy drilling equipment to operate. With average temperatures below −50 °C pack ice can be reinforced by pumping up seawater from depth and spraying it on the ice, when it freezes immediately. As the summer approaches wells are capped and installations removed by air to await reconnection next season.

These operations involve physical disruption of the environment and all the ecological disturbance which inevitably accompanies invasion by numbers of human beings. There is also the ever-present hazard of a major oil leak or accidental discharge of drilling mud, containing high concentrations of toxic cobalt and chromium salts. These pollutants, imprisoned in the ice, may not be dispersed by the usual agents of wind, sea, and run-off. Biodegradation of oil renders it harmless in time – a temperate shore contaminated with spilt oil returns to almost pristine condition in ten years – but the microbial activity which achieves this is slowed down at low temperatures. Precautions against these risks are taken and contingency plans in case they occur are drawn up but, nevertheless, usually because of human frailty, catastrophe strikes. A notorious example is the collision with a well-charted rock in Prince William Sound (60°30′N 147°E) in March 1989 of the *Exxon Valdez*, a fully loaded 300 m tanker. The 38-million-litre spill formed a slick covering 2500 km^2 of the Sound, described in 1973 as 'a resplendent . . . natural wealth as yet unassailed by unmanaged human use'. It is an enclosed sea area, sub-Arctic rather than Arctic in nature, within which the oil terminal at Port Valdez is sited. An ecological survey, conducted before the terminal was constructed, to provide a basis for assessing the effects of release of treated tanker ballast water, concluded that the impact would be negligible. The response to the *Exxon Valdez* spill was disorganized and decisions were delayed. Within two weeks the carcasses of 150 otters and over 1200 birds, including 146 bald eagles, were counted – and this was a small fraction of the actual kill. One estimate is that half of the pre-kill population of birds were killed – 490 km of shore were cleaned but, in order to complete this before winter, high pressure hot water sprayers were used, causing almost as much ecological havoc as the oil itself. The expense of the spill to the oil company was in excess of US$2 billion.

Oil and gas extraction has also been carried out on a large scale in Arctic Russia at similar hazard to the environment. Leaks of crude oil from an ill-maintained pipeline near Usinsk (65°57′N 57°27′E) in 1994 seem to have amounted to three times the *Exxon Valdez* spill. Fortunately, most of the migratory birds had gone south before the spill. Makeshift earthen dykes did not prevent its escape to threaten the delta of the Pechora River, a national park. Fisheries in the region declined from a catch of 68 tonnes in 1976 to 8.4 tonnes in 1994. The tundra over an area of 70 km^2 was seriously damaged and herding of reindeer disrupted (Stone 1995).

Although the Antarctic is not yet oil-producing, it is still subject to oil leakages from ships. The sinking of the Argentine supply ship *Bahia Paraiso* about 1.6 km from the US Palmer Station (64°46′S 64°03′W) in January

1989 illustrates what may happen in the frigid marine environment. More than 680 000 litres of diesel fuel were released, covering an area of some 100 km^2 of sea within four days. Mortality amongst littoral invertebrates was obvious but the greatest impact was on seabirds, although, because most of the affected population moved out to sea shortly after the accident, it was not possible to estimate the numbers killed. The US National Science Foundation responded within 36 hours – research at Palmer was in jeopardy – sending boats and equipment to control the damage and following this up with a programme of impact assessment. Hydrocarbon concentrations in the Antarctic marine environment generally are low and the compounds represented are overwhelmingly of biogenic origin, pollution being local and confined to few events. The Antarctic provides a yardstick against which global pollution by hydrocarbons can be judged (Cripps and Priddle 1991).

Sanctioning and regulation of mining and oil and gas extraction in the Arctic are matters for national governments, which, on the whole, have been anxious to promote such development. The situation is different in the Antarctic, where the concept of national sovereignty has been under eclipse since the advent of the Antarctic Treaty and governments have had different objectives in view than commercial development (p. 235; Walton in Ives and Sugden 1995). Considerable reserves of minerals, oil, and gas are probably present in Antarctica. Parts of it, notably the Peninsula, are geologically related to some of the richest metal-producing areas in the world and copper, iron, molybdenum, lead, zinc, and silver deposits have been located. Coal is there in abundance and indications of natural gas and oil have been found. So far, however, there is nothing worth commercial consideration but, given that only one per cent of the land surface is accessible to direct geological scrutiny, the probability of there being rich deposits somewhere on the continent is great. If these were found, the cost of extraction would be extremely high; the Alaskan oil fields are only just economically viable and the difficulties which mining or oil drilling would face in Antarctica are far greater. Antarctic scientists have been confident that there is no immediate threat of exploitation but nevertheless the Scientific Committee on Antarctic Research (SCAR), which advises the Antarctic Treaty organization, drew up a convention for regulation, which was agreed in 1988. This would have permitted exploitation of mineral and oil resources under regulations so stringent that it would be unlikely to be attempted in the foreseeable future. This did not satisfy the conservationist lobby, which wanted nothing less than total prohibition. France and Australia, two of the signatories, had second thoughts and supported this view. As a result, in 1991, a moritorium for at least 50 years on all exploitation of minerals and oil in the Treaty area was agreed. This is to be welcomed with reservations. The abandonment for political reasons of a carefully constructed and scientifically based convention does not bode well. Not all nations subscribe to the Antarctic Treaty and international agreements have been known to be renounced. The discovery of an exceptionally rich reserve might lead to unscrupulous exploitation with no precautionary regulations in place to be invoked.

9.9 Pollution

Pollution is defined by Holdgate as 'The introduction by man into the environment of substances or energy liable to cause hazards to human health, harm to living resources and ecological systems, damage to structures or amenity, or interference with legitimate uses of the environment'. The Arctic, when occupied only by its aboriginals, and the Antarctic until 200 years ago, were unpolluted. Even now, so extensive are the areas and so small the number of humans in them, that it might be thought that pollution should still be negligible. This is not so because, in the first place, access points into the polar regions are few and tend to be where plant life and animal congregations are greatest so that the impact of pollution can be considerable. Second, much pollution, carried by air or water, is now global and affects polar regions as well as everywhere else. Indeed, in the Arctic, circumstances conspire to concentrate contaminants. The rivers flowing into the Arctic basin collect materials, including heavy metals and organochlorine pesticides, from catchment areas in which mining, forestry, and agriculture, are locally intensive. These materials may be taken up by plankton (*bioaccumulation*) and be concentrated as they are passed along the food chain (*biomagnification*). The particulate organic matter containing these pollutants may itself be concentrated at fronts (p. 24) and ice edges (p. 174) peripheral to the Arctic basin. It is not surprising, therefore, that polar bears feeding along the ice edge in the Svalbard region contain high concentrations of mercury, cadmium, and organochlorine compounds such as DDT derivatives and polychlorinated biphenyls (PCBs) (Davis 1996). On the other hand, indiscrimate dumping by the former USSR of radioactive waste in the Kara Sea seems likely, in view of the low solubility of the potentially dangerous substances in cold water, to produce only local contamination and not be too serious a menace.

Pollution takes many forms, some of which have already been discussed (pp. 20, 115). No doubt there are unrecognized pollutants too. Thus, only recently has it been realized that radiations from power networks cause precipitation of high energy particles in the upper atmosphere, producing perturbations of navigational signals. This surely falls within the definition given above although its biological consequences are perhaps remote. Pollution around scientific stations has attracted particular attention. Sited at an abandoned coal-mining settlement at Ny Ålesund, Spitsbergen (78°55′N 12°00′E), the international research centre which includes Norwegian, German, and British polar laboratories, perhaps has an advantage in starting from a low level of environmental cleanliness. In the Antarctic, the US McMurdo Station has had a name for mess and pollution. A survey of soils in its vicinity showed contamination with lead, zinc, and copper, not constituting serious pollution but indicating definite changes in chemistry due to human activities. Stations belonging to other nations are on a smaller scale and are lesser offenders. Marine and shoreline sediments up to 1 km from the Australian station, Davis, Prydz Bay (68°35′S 77°58′E) contain appreciable amounts of the faecal sterol, coprostanol, emanating from the sewage outfall. That this was significantly lower than in corresponding sediments near McMurdo is attributed to the latter discharging untreated sewage whereas that from Davis receives secondary treatment. Degraded hydrocarbons derived from fuel spillage and vehicle usage occur in soils around Davis. Such data provide a baseline against which the effectiveness of pollution control can be assessed. Most nations involved in Antarctic research now make efforts to clean up their stations and reduce pollution. Both the United States and Australia have introduced legislation to these ends and New Zealand and the United Kingdom ensure that both scientific staff and visitors are aware of the need to avoid pollution. The Environmental Protocol adopted by SCAR in 1991 has the potentiality of maintaining Antarctica in essentially pristine cleanliness – except, that is, for global pollutants brought via the atmosphere or sea.

Arctic haze is a visible sign of atmospheric pollution. When atmospheric inversions (p. 6) occur, a haze, originating in pollution carried from cities and industrial installations further south, reduces incident solar radiation by as much as 15%. Oxides of sulphur and nitrogen are components in this pollution, as in 'acid rain', and can have marked effects on the weathering of rocks as well as on vegetation. Already, effects on climate may be beginning (p. 243) but there do not yet seem to be any serious effects on organisms. The first step in preventing further deterioration must be clarification of international legal responsibilities. The more insidious pollution by 'greenhouse gases' and CFCs have already been discussed (pp. 20, 46). The discovery of DDT in penguins and a crabeater seal from the Antarctic first drew attention to the world-wide distribution of this stable pesticide. This, and related organochlorine substances, such as PCBs, have subsequently been found both in the Arctic and Antarctic, in organisms such as phytoplankton, zooplankton, fish, mosses, and lichens as well as in snow and freshwater and marine deposits. These pollutants are transported via the atmosphere. Their volatility is low but tropical climates favour their volatilization whereas polar cold favours deposition from vapour in the air – they are, in effect, distilled from one place to another. In Southern Ocean water the concentration of PCBs is only a seventh of that in northern temperate zones but because of bioaccumulation their concentrations in particulate matter are as high in Antarctic waters as elsewhere. The concentrations found do not, however, seem seriously to affect plant and animal life.

In the wider field of the understanding and management of pollution the polar regions are pre-eminent for the information they provide on the past course and present trends of global contamination. As snow falls it brings down particulate matter from the air and as it consolidates into ice it encloses bubbles of the air itself. A core of polar snow and ice thus comes to contain a record in the form of samples stored in chronological order which can tell us much about the atmosphere in the past. Except locally, near centres of human activity, the record is global and similar in essentials whether it comes from the Arctic or Antarctic (Wolff 1990; Lorius 1991; Oerschger in Hempel 1994). Analysis of the trapped air of ice cores shows that its carbon dioxide content has increased by 25% in the last 200 years and that of methane has more than doubled. There is a close correspondence between the concentrations of these 'greenhouse gases' and temperature, deduced from isotopic ratios, during the last glacial cycle back to 160 000 BP. This is, of course, a correlation rather than a proven causal relationship but, taken with the present steady rise in atmospheric carbon dioxide concentrations it leads us to expect a period of global warming. Mathematical models suggest that the warming will be greatest in the Arctic and least in the Antarctic, which is stabilized by the Circumpolar Current. However, there are factors, most importantly, apart from the uptake of carbon dioxide by the oceans, the masking effects of aerosols produced by industrial pollution, which work against this. There is no convincing evidence yet but it is possible that the effects of one form of pollution may be cancelled out by the effects of another.

Profiles of radioactivity in snow record testing of nuclear bombs in the atmosphere and also accidents such as that at Chernobyl. Nitrate has more than doubled and sulphate more than trebled in concentration in Greenland snow, corresponding to increase in haze, but in Antarctica, further from industrial activity, there have been no significant rises. Lead increased a 100-fold up to the 1970s in the Arctic but is now declining whereas the increase in Antarctica has been slight. It is important that local pollution should be minimized, especially in Antarctica, if snow and ice are to continue to provide a useful index of the state of our atmosphere.

9.10 Tourists and other recreational visitors

A visit to polar regions is an experience for which people will pay. As long ago as 1896 there was a hotel on Spitsbergen catering for tourist ships. In 1966, Lars-Eric Lindblad set a pattern of cruises to the Antarctic, combining comfort with instruction and a sense of adventure. Now, increasing numbers of cruise ships of varying degrees of luxury and increasingly larger size visit both polar regions. They have been taken through the Northwest and Northeast Passages, to the North Pole, around Antarctica, down the west coast of the Peninsula, and into the Ross Sea. Recently, the availability of Russian ice-breakers for charter has allowed particularly venturesome expeditions. High powered inflatable boats provide an efficient means of landing on beaches and helicopters can fly visitors inland. In the 1992–3 season 6983 tourists entered the Antarctic Treaty area aboard ships, 54 on yachts, and 185 on aircraft (Enzenbacher 1994). Day flights to Antarctica from New Zealand, discontinued after a crash on Mount Erebus in 1979 killed all on board, allowed about 10 000 people to see Antarctica. Flights have now been resumed, from Australia.

Overflights obviously have minimum impact on the environment. Cruise ships provide for a potentially environmentally satisfactory kind of tourism which hotels and resorts do not. Sea-passengers, mostly educated and highly motivated towards conservation anyway, can be kept under control but, nevertheless, their impact is concentrated on landing places with abundant wildlife (Fig. 9.1). Captains should have information about reserves and sites of special scientific interest and avoid them but this does not always happen (p. 128). The International Association of Antarctica Tour Operators (IAATO), formed in 1991, is a means of ensuring standards but not all tour operators belong to it. In Antarctica the main attractions lie in scenery and wildlife, although climbing the Vinson Massif (78°02′S 22°00′W) is becoming popular with those who can afford the air fare. In the Arctic there are the extra interests of the indigenous peoples, hunting and fishing, and the use of float-planes or dirt airstrips allows visitors to penetrate deep into remote areas. The possibilities of environmental damage increase accordingly, especially if there is indiscriminate use of land vehicles (Fig. 9.2). No one knows what the effect of the shattering roar of the snowmobile on the fauna of the polar wilderness may be.

Fig. 9.1 Tourists photographing wildlife, Deception Island. The rule that seals and birds should not be approached nearer than 5 metres is being observed, approximately. (Photo G. E. Fogg.)

Some research has been done on tourist impact on polar habitats and wildlife (Stonehouse 1992). In Antarctica this has concentrated on penguins. The numbers of nesting pairs of Adélie penguins in a rookery near McMurdo Station (Cape Royds) declined to about a half of what they were before the station was established and disturbance by helicopters and visitors began. Following remedial measures there was recovery (Fig. 9.3). The effects of occasional small parties of tourists on foot are debatable. Certainly the birds are the first to ignore the rule that the distance between humans and them should never be less than 5 metres and give the impression of

Fig. 9.2 Impact of large-tracked vehicles on soil and hydrology; Fildes Peninsula, King George Island, South Shetlands, within the area of a former Soviet biomonitoring site. (From Harris 1991, by courtesy of the author and the editor of *Polar Record*.)

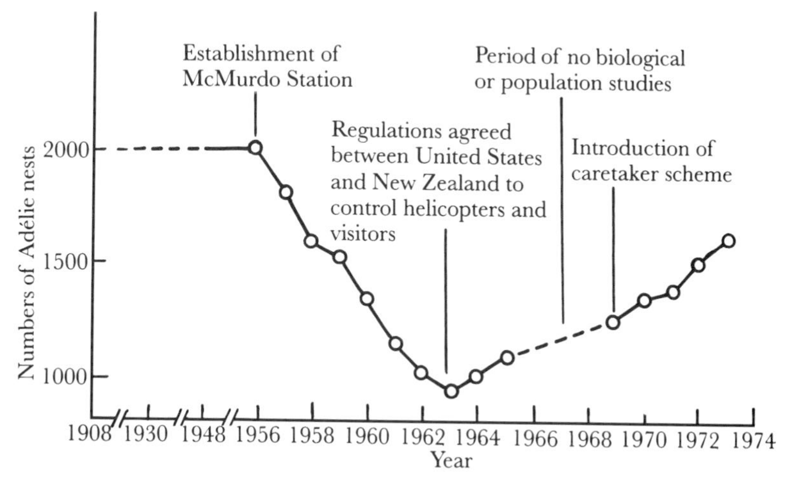

Fig. 9.3 The effect of disruption by humans on the Cape Royds (77°33′S 166°08′E) Adélie penguin rookery and its recovery after remedial measures were taken. (Courtesy of R. B. Thomson.)

curiosity rather than upset. Attempts have been made to get a quantitative measure of stress by electronic monitoring of heart rate. Rates increased by an average of 19% on approach by a human but this was in birds stressed by fitting of the recording device and hence predisposed to alarm on sighting people. One is inclined to the flippant comment that one's own heart rate may go up on the approach of a human being but this is not always an indication of stress. Analysis of remote-control video records of bird's reactions when tourists are about may provide more reliable evidence. Tourism in polar regions is set to increase in the future and we certainly need to know more about its impact.

9.11 Conservation

Conservation means different things to different people and has come to arouse angry passions but here it will be taken as the wise use of resources rather than the not-so-simple preservation of the *status quo*. Management for conservation must be based on sound science but whether this happens depends on expediency, politics, or, sometimes, fanaticism. Approaches and means of achieving conservation are different in the two polar regions. It will be best to discuss the reasons for this before proceeding to detail.

9.11.1 The conservation politics of the two polar regions

Ownership of Arctic territories has been settled for nearly a century with every portion of land north of the Arctic Circle allocated to one of eight nations: Canada, United States, Iceland, Denmark, Norway, Sweden, Finland, and the former USSR/Russia. Each country uses its land as it sees fit and can support resource management by law. Systems have, however, been different. That of the the former USSR, for example, was decided by diktat, heavily resource orientated, and restricted by security considerations. By contrast, resource utilization in Alaska has been decided democratically with public debate, voluminous reports, lobbying by industrialists and conservationists, congressional decisions, and lawsuits. There has been some collaboration between nations. International legislation covers cross-frontier movements of reindeer. The exploitation of marine resources, once free for all, is controlled by the 1982 Convention on the Law of the Sea with 12-nautical mile (22 km) territorial limits and Exclusive Economic Zones. Arctic research has largely depended on private enterprise and reflected the jigsaw pattern of sovereignty with little co-ordination between the programmes of different nations (Levere 1993). The International Biological Programme of 1964–74 achieved some standardization of methods and exchange of ideas in tundra research (Rosswall and Heal 1975). The recently established International Arctic Science Committee may in the course of time provide a general framework for research and conservation.

Several nations have claimed parts of Antarctica. First, Britain in 1908 claimed the sector between 20°W and 80°W with its apex at the pole with the conservationist objective of regulating the whaling industry. Further claims by Britain and other nations, however, were frankly imperialistic and led to much diplomatic bickering and occasional exchange of shots in the decade following the Second World War. The United States and former USSR neither made claims nor recognized those of others, but the United States used Antarctic waters for major naval exercises. The situation became tense but was unexpectedly resolved by the International Geophysical Year of 1957–8. Apart from being scientifically successful this demonstrated that nations of opposed ideologies could collaborate amicably in the pursuit of a common purpose in a hostile environment. Diplomatists, to their credit, seized on this and the Antarctic Treaty, putting claims of sovereignty into cold storage and dedicating the region to science was signed in 1961 (Triggs 1987; Fogg 1992). The initial number of parties was twelve but some forty nations are now involved, representing about four-fifths of the world's population. Decisions are reached with the informal advice of SCAR and do not become effective until ratified by all the Consultative Parties. There are requirements that Consultative Parties should be actively involved in research in Antarctica and that they should co-operate and share information among themselves. There is thus no incentive for individual states to direct their research towards strategic or commercial ends and scientists have been virtually free to follow their own interests. As frequently happens when this is allowed, some of their results have been both unexpected and of great significance for mankind, particularly in connection with the management of the global environment. The Treaty organization has been criticized as an exclusive club and there has been pressure from non-member states for Antarctic affairs to become the responsibility of the United Nations, a move which is supported by several conservation bodies. This seems undesirable since if science is to be the main business it is necessary that decisions should be taken by those with knowledge and practical experience and not in open debate in which ideology would inevitably prevail. In any case, the record of the United Nations in environmental matters has not been impressive.

9.11.2 The Arctic

The Arctic is resource-rich but in the absence of local markets there have to be extensive transport systems to move personnel and products, all involving damage and pollution. The multinational organizations which exploit these resources naturally put economic considerations first and have no compelling reason to be concerned for the polar environment. The region is also of strategic importance and it seems unlikely that the United States and Russia will appreciably reduce the extent of their military installations for the present. The use by the former USSR of Novaya Zemla for nuclear weapon testing and the dumping of radioactive waste in the Kara Sea have produced environmental problems with international repercussions. On the positive side, all the sovereign states of the Arctic have established national parks, scientific or scenic reserves, and wilderness areas. In Russia, for example, the area of all types of Arctic reserve amounts to 19.7 million hectares, about 10.2% of the total. They include five strict nature reserves with a complete ban on economic and other activities such as eco-tourism. Unfortunately, these strict reserves do not encompass all the different types of Arctic habitats. Reserves are well staffed but poorly financed and the legal framework is not always certain, much depending on presidential decrees that may have limited effective lifetimes (Volkov and De Korte 1994). In all countries there is pressure from commercial interests for concessions which can usually be plausibly shown to have little immediate environmental impact but which, if granted, may have unforeseen ecological consequences or set a precedent for further encroachments which cumulatively and imperceptibly destroy the value of a reserve.

9.11.3 The Antarctic

Few of the explorers or the scientists who followed them had much concern for conservation and only brief reference was made to the necessity for preservation and conservation of living resources in the Antarctic Treaty. However, any sort of military activity, weapon testing, nuclear explosions, and disposal of radioactive waste were prohibited in the Treaty area and this prohibition, which has been strictly observed, disposed at a stroke of some of the worst forms of environmental destruction and pollution. The need for being more comprehensive and explicit about conservation was soon realized by SCAR and 'Agreed measures for the conservation of Antarctic fauna and flora' were formulated by the Treaty organization in 1964. These enjoined governments to prohibit the killing of animals save for scientific purposes or in cases of necessity, to minimize disturbance of bird and seal colonies, to prohibit the collection of plants except for scientific purposes, and to alleviate pollution. Areas of outstanding scientific interest were to be protected. Around the same time the convention to protect seals was agreed. In 1970, there were further recommendations on the use of radioisotopes, tourism, non-governmental expeditions, and collection of data on flora and fauna. Strictly speaking, the Treaty area does not cover the high seas but an ambitious exercise based on the concept of the marine ecosystem as a functional whole was embodied in the Convention on the Conservation of Antarctic Marine Living Resources. A moritorium on mineral and oil exploitation was agreed (p. 228). These measures have been remarkable in that, mostly, they have been based on sound scientific principles and produced in advance of serious damage being done. In practice, their application has been somewhat uneven – the failure to protect fish stocks and the disastrous breakdown of conservation measures on King George Island, where nine nations have scientific stations (Fig. 9.2), are cases in point – and there has been severe criticism from the more aggressive conservationists. One has to agree that as the 'last unspoilt continent', Antarctica needs special protection. Nevertheless, conservation in Antarctica is far in advance in organization and effectiveness than anywhere else. With the setting up by SCAR of a Group of Specialists on Antarctic Environmental Affairs and Conservation and the progressive enforcement of the Environmental Protocol of 1991, it should greatly improve. There must always be a sensible balance between conservation and use. The idea of Antarctica becoming a world wilderness park, in which no habitat or living organism may be interfered with, might severely restrict research which gives us insights, unobtainable from any other part of the world, alerting us to previously unrecognized dangers and essential for the management of the global environment (Cook 1990; Hempel 1994).

10 Some conclusions

The quintessential polar habitat, the ice cap, supports no vital activity but does not extinguish life altogether; some microorganisms can lie dormant in it for seemingly indefinite periods. Lack of moisture, not low temperature itself, is the ultimate constraint and wherever liquid water appears, active life becomes possible and is usually to be found. Around the ice cap such situations are transient and shifting, being colonized only by such opportunistic organisms as are already present in dormant form or whose propagules arrive via air, water, or human visitors.

Organisms establishing themselves in terrestrial habitats at the ice margin must be able to grow and reproduce in the short intervals when conditions are favourable and to survive desiccation, cold, and starvation when they are not. Among them there need to be autotrophs, usually algae or cyanobacteria, but often there is some form of structured association with heterotrophs. The associations found in lichens, algal mats, endolithic communities, and in the brine channels in sea ice are more successful in utilizing the resources available in these extreme habitats than are individual components alone. In the endolithic and brine channel communities this is evidently because the combination of phototrophs and heterotrophs allows existence in an almost self-contained microcosm, within which materials can be cycled and conserved. Lichens are more effective in absorbing and retaining water than are the algae they contain. Such associations may support a limited microfauna of detritivores but their productivity is minimal and they do not provide for much secondary production.

Thus far, the Arctic and Antarctic are similar in the habitats they provide and in the species which occupy them. In peri-glacial terrestrial areas, however, there are differences. These stem from the respective degrees of isolation from the rest of the biosphere, the times for which they have been open to colonization, the extents of the land surfaces, and the different climates. In both regions it seems that none of the organisms found have any unique adaptations to polar life but have migrated from milder climates already genetically equipped with physiology and life cycles which can be modified for survival in the harsher environment. Accumulation of biomass, as in Arctic tundra, is sometimes sufficient to ameliorate the environment. Despite the longer time that has been available for colonization, the greater distances and lack of bridges facing migrants into the Antarctic have resulted in floras and faunas much poorer in species than those in the Arctic and, conspicuously, lacking in mammals. With the Arctic mammals there is, again, no fundamentally different feature of physiological adaptation, survival being largely a matter of modification of life cycle and behaviour. This is particularly evident in the Inuit, who have no special structural or metabolic adaptations beyond the level of acclimation for dealing with cold. They create their own microenvironments by effective clothing and housing and avoidance of exposure. The secret of success of polar organisms has been neatly summed up as a combination of an ability to 'sit it out' when conditions are adverse with a 'get up and go' strategy in the short favourable periods. Neither of these are exclusively polar characteristics. Nevertheless, the way in which structural, physiological, biochemical, and behavioural features are combined, as in the emperor penguin, to enable a species to be successful in a quite atrocious environment is something which one cannot dismiss as fully explained.

The contrasts between terrestrial and marine polar habitats could scarcely be greater. In place of desiccation, temperatures which can fluctuate between extremely low and uncomfortably hot, and exposure to destructive winds, the open sea offers unlimited access to liquid water, stable temperatures never falling below freezing point, and currents which are not usually life-threatening. Habitats in ice-free polar seas or on their shores show no features differentiating them in any fundamental way from temperate waters. Although the Southern Ocean has often been described, superficially, as having the simplest of food chains – diatoms, krill, and whales – the basic trophic structure is the same in polar seas as in temperate ones. The complexities of the food web, particularly at the level where the ultraplankton meshes with the higher trophic system, seem just as great. The sea has fronts, adverse currents, and deep basins as barriers to invading organisms. However, in the Arctic, convoluted shorelines and current patterns, together with extensive and shallow continental shelves facilitate invasion. Organisms from temperate waters may penetrate far north and it is not always easy to distinguish between Arctic and sub-Arctic floras and faunas. In contrast, the northward edge of the Southern Ocean is marked clearly, biologically as well as physically, by the Antarctic Polar Front and its currents deflect water-borne invasions. For some organisms the Front is an almost impermeable barrier. The 25 million years which have elapsed since it was established have allowed the evolution of distinctive biotas with high proportions of endemic species. Among these, the fishes have gone furthest in developing physiological and biochemical mechanisms for survival under polar conditions.

Ice gives polar seas their unique habitats. Its immediate effects are mostly physical – abrasion, modification of radiation flux, of temperature, of stability of the water column, refuge from predators, and a solid substratum for birds and seals above and microalgae below. The microhabitats provided by the brine channels in ice are inhabited by a microbial community essentially similar in trophic organization to the endolithic communities of the polar desert but lasting for a season or two rather than for centuries. Another difference is that whereas the organic production of the endolithic community finds its way into general circulation only slowly, that of the sea ice community is immediately available to secondary producers at ice melt. The existence of this niche, depending essentially on the pattern of ice crystallization, in the Southern Ocean but not in the Arctic Ocean, is another instance of the far-reaching effects of the enclosure of the Arctic Ocean by land masses. The ice edge, a distinct and remarkable habitat which shifts hundreds of kilometres polewards each summer, is highly dynamic, both physically and biologically. The surge in primary and secondary production which follows it is adequately accounted for in terms of release of organic matter from the ice and stabilization of the water column providing conditions favourable for phytoplankton. This is an intensification of what happens in the open sea, not anything radically different.

Primary productivity in polar habitats is generally less on an area basis than that of equivalent vegetation elsewhere, although in favoured situations it sometimes equals or even exceeds them. As everywhere, the primary productivity of submerged aquatic communities is less than that of terrestrial ecosystems with adequate water. The limitation on photosynthesis is not temperature *per se* but is usually desiccation or nutrient supply on land and nutrient deficiency or excessive turbulence in the sea. There is no specifically polar factor preventing plants from doing better than they do.

Polar ecosystems are often described as 'fragile'. This requires some qualification. Habitats may be obviously fragile. Frost heaving, wind, and ice movement can wreak havoc, and activities of humans, or even of the native fauna, may have devastating effects. Traffic around a scientific or military installation cuts up the ground and its scars persist for decades. On the other hand, human impact on the ice is minimal and disturbance is obliterated within a season. Equally, polar communities show extremes of fragility and robustness. The macroscopic vegetation of polar desert is usually physically fragile, if biologically robust, and a footstep may destroy a lichen a century or more old. Wet tundra does not invite passage either on foot or in a vehicle and its vegetation is more resilient. However, it is sensitive to alteration in drainage or heat balance and an oil spill, by altering albedo, excluding oxygen and introducing both organic substrates for microbial growth and toxic materials, can have profound effects. The character of freshwaters can be radically changed (e.g. switched from an oligotrophic to eutrophic state), especially if the volume is small, by quite minor events such as interference with an inflow or introduction of small amounts of inorganic or organic substances. At the higher trophic levels, hunting or fishing may upset the balance of predator and prey. Where a single animal occupies a key position in a terrestrial ecosystem, as the lemming may in Arctic tundra, reduction in numbers of its predators can lead to rapid population increase with drastic effects on the structure and composition of the vegetation.

Nevertheless, many plants are robust and recovery may take place even if it is a lengthy process. Marine ecosystems are resilient. The severe reduction in whale numbers in a few decades cannot really be cited as evidence of fragility since the biomass of these top predators is a minute part of the total and their place can be taken by other predators with no perceptible effect on lower trophic levels. Overfishing of krill, which occupies a key position in the Southern Ocean akin to that of the lemming in the tundra, might have far-reaching effects. Simple mathematical models suggest that if krill were harvested at a rate equal to half the population's intrinsic growth rate (which would give maximum yield), the numbers of top predators such as whales and seals would fall rapidly to new equilibrium values (May *et al.* 1979). It seems probable that the repercussions of even catastrophic changes at these top trophic levels would be imperceptible at lower levels. The relations between the ultra- and microplanktonic communities seem much the same in all parts of the oceans, regardless of what goes on up top. Microorganisms are much less susceptible to the toxic effects of oil and organochlorine pollutants than are the higher animals. Zoologists and microbiologists often have quite different notions of fragility or absence of life.

Investigations of polar ecology have made notable contributions to biological science in general, principally because some polar habitats present situations in which particular problems can be studied under unique conditions. The total absence of mammalian herbivores and the small number of species generally in terrestrial habitats in the Antarctic results in food webs, flows of energy, and cycles of materials of greater simplicity than those elsewhere. They are consequently more amenable to mathematical modelling. Large, accessible, undisturbed colonies of seals and seabirds adjacent to sea areas which are under investigation by marine biologists give unrivalled opportunity of investigating the role of these animals in the marine ecosystem. The tolerance to handling of Antarctic seabirds has helped quantitative studies on diet and growth as well as simplifying the deployment of telemetric and other electronic monitoring equipment. Polar ornithologists have led the way in studies of the ecology of seabirds. The habits of the Weddell seal allow its diving patterns and physiology to be studied under nearly natural conditions using techniques usually possible only in the laboratory. As a result, the diving behaviour of this seal is better known than that of any other marine mammal. Apart from allowing innovative long-term studies of microclimate and microbiology, the dry valleys of Antarctica have provided the nearest approximation on earth to the surface of Mars, a terrain for testing ideas and equipment used in searching for evidence of life on that planet.

The study of polar habitats has other wide perspectives. For various reasons – because the explorers and scientists who first ventured into these regions were men with broad interests and because it has always been practical to send naturalists together with astronomers, magneticians, and geologists on expeditions – polar science has always been holistic in outlook. In contrast to the reductionist approach, which studies processes in simplified situations under controlled conditions, the holistic view recognizes that the manifold processes taking place in the natural world interact with one another to varying degrees and that conditions are never constant. Reductionist science is essential but it is not always realistic to apply it unthinkingly in the natural environment, with all its complexity and inconstancy, and to do so may be disastrously misleading. Study of natural systems as a whole has hitherto been largely empirical but is now acquiring rigour by use of *in situ* experiments and mathematical modelling.

From the holistic viewpoint, the polar regions cannot be put aside as remote, separate, and distinct from the rest of the world. The Arctic and Antarctic, because they are our planet's heat sinks, play key roles in propelling flows of air and water, setting the patterns of circulation in atmosphere and ocean, influencing our weather, shipping, and air transport. At the biological level, the polar seas act as sinks for carbon dioxide, counteracting to some extent the man-made 'greenhouse effect'. The copious release of dimethylsulphide by algae in polar waters may play a part in determining cloud cover, thus affecting the earth's heat balance in another way. The isolation and physical conditions of the Arctic and Antarctic enable global perturbations of atmospheric chemistry (e.g. in ozone and pollutants), to be detected more certainly there than anywhere else.

To understand the present situation it is well to consider what has happened in the past. Over many millions of years the polar regions have undergone major changes in configuration of their land masses and climate with corresponding transformation of habitats. However, it is the geologically brief period beginning around 120 000 BP, including a warm interglacial followed by the Würm glaciation and then the interglacial in which we now live, which is of most relevance here. Study of fossils, of beetles in particular, in deposits laid down in glaciated areas have shown that conditions have never been settled for very long and that change can be rapid. Physicochemical examination of cores of polar ice and sediments confirms this picture and yields detailed records of the chronology and magnitude of glaciations and global warmings. It appears that the last ice age ended abruptly, perhaps within the span of a human life (Dowdeswell and White in Wadhams *et al.* 1995). These changes cannot be put down to any human agency. Now, man's impact on the environment is mounting. We have no reason to suppose that the present state which we take to be normal is any more stable than those in the past and by our activities we are destabilizing the global system and making change more likely. The totally unexpected development within a decade of the 'ozone hole' over Antarctica makes this abundantly clear.

In this situation it is only sensible to assemble information about the functioning of the global environment and to see whether any predictions of its future may be made (Drewry *et al.* 1992; Hempel 1994; Wadhams *et al.* 1995). The numerical models which are being used to do this have to handle vast amounts of data and take complex interactions into account. The most recent models require as many as 2×10^{14} numerical steps. Even so, some factors which are undoubtedly crucial and which involve polar ecology have not yet been included. Current models agree in indicating that global warming will take place but it cannot yet be said with any degree of certainty that disastrous changes of climate and sea level will happen in the foreseeable future. Nevertheless, there is some probability of a 2.5 °C rise in global mean temperature by the middle of the 21st century. This rate of warming is 10–100 times faster than the rise of approximately 5 °C which took place between 15 000 and 10 000 years ago.

A complementary approach is to monitor changes in the environment. Observed rises in sea temperature, retreat of ice fields, and a decrease in depth of mixing in the Greenland Sea, point to global warming being underway but these changes still lie within the limits of 'noise' in the system and give no certain information yet. It can be argued that polar ecosystems are more sensitive and reliable indicators of change than these physical measurements. Changes in climate are predicted to be greater and more rapid around the poles and polar species are particularly vulnerable to change because of their slow growth and low fecundity. They are especially responsive to rise in temperature since they exist so close to the lower temperature limit for survival (Callaghan *et al.* in Drewry *et al.* 1992; Callaghan and Jonasson in Wadhams *et al.* 1995). Here again, there are differences between Arctic and Antarctic. Models agree in predicting that the Arctic is particularly sensitive. A doubling of atmospheric carbon dioxide concentration, for example, is likely to cause a mean rise of 4 °C over the whole year in the Arctic as compared with 1.7 °C averaged over the whole earth and a negligible rise in the Antarctic. However, information on responses of Arctic plants and animals in their natural habitats to supposed climate changes is, as yet, sparse. In the Antarctic the predominantly cryptogamic vegetation is not markedly responsive to temperature change and barriers to population shifts are formidable. But there is some suggestion that recent rapid population expansions of the two indigenous flowering plants are related to climatic amelioration.

The impacts of global warming on polar habitats and communities are complicated and difficult to predict (Oechel *et al.* in Woodin and Marquiss 1997). The response of individual species to continuous and rapid climate change is generally one of migration rather than of evolutionary adaptation (Huntley 1991). This seems to have been true in the Arctic and, discounting notothenioid fish and emperor penguins which have had many millions of years to adapt to change, in the Antarctic too. Species differ in the timing, rates, and directions of their migrations so that communities must be regarded, not as being of fixed composition, but as temporary assemblages of species, associating and dissociating as conditions alter. Rapid change will thus produce communities different from those arising from slower change. The maximum migration rate of most trees is between 150 and 500 m yr^{-1}, this being just about enough to have kept pace with retreating ice at the end of the ice age. The present rate of warming being so much faster than it was then, we can no longer expect forest to follow retreating ice. Migration rates of pests, pathogens, or herbivores may be different from those of invading plant species and so upset ecological balance. Perhaps the greatest immediate effect on the biota will follow the disappearance of permafrost, which may lead either to waterlogging or to drought according to topography and precipitation regimes. Warming will accelerate the decomposition of the substantial organic carbon reserves in tundra, probably turning them into net producers of 'greenhouse gases' and thus providing positive feedback to global warming. In the Antarctic there are no great areas of tundra underlaid by permafrost to provide feedback. Attempts to manage terrestrial ecosystems to avoid unwanted effects of global warming in the polar regions will tax our ecological expertise to its limits.

Effects on the polar seas will be of a different kind. It seems unlikely that a rise in temperature will affect total primary productivity and changes in secondary production seem unpredictable. There may, however, be considerable alterations in the species composition and distribution of marine communities, including seabirds and fishermen, in response to rise in sea level and shifting patterns of water movement (Fogg 1991*b*).

Polar habitats are unique and of great intrinsic interest to ecologists. Their study also promises to help us understand, and to some extent cope with, the damage to the global environment which we have so unthinkingly wrought.

Further reading

Battaglia, B., Valencia, J., and Walton, D. W. H. (ed.) (1997). *Antarctic communities: species, structure and survival.* Cambridge University Press.

Chernov, Yu. I. (1985). *The living tundra* (trans. D. Love). Cambridge University Press.

Crawford, R. M. M. (1995). Plant survival in the High Arctic. *Biologist*, **42**, 101–5.

Davenport, J. (1992). *Animal life at low temperature.* Chapman & Hall, London.

Drewry, D. J., Laws, R. M., and Pyle, J. A. (ed.) (1992). Antarctica and environmental change. *Philosphical Transactions of the Royal Society of London*, **B338**, 199–334.

Eastman, J. T. (1993). *Antarctic fish biology: evolution in a unique environment.* Academic Press, San Diego.

Fothergill, A. (1993). *Life in the freezer: a natural history of the Antarctic.* BBC Books, London.

Green, B. (1995). *Water, ice and stone: science and memory on the Antarctic lakes.* Harmony Books, New York.

Hardy, A. (1967). *Great waters.* Harper & Row, New York.

Hempel, G. (ed.) (1994). *Antarctic science: global concerns.* Springer, Berlin.

Ives, J. D. and Barry, R. G. (ed.) (1974). *Arctic and alpine environments.* Methuen, London.

Ives, J. D. and Sugden, D. (ed.) (1995). *Polar regions.* RD Press, Surrey Hills, NSW, Australia.

Knox, G. A. (1994). *The biology of the Southern Ocean.* Cambridge University Press.

Laws, R. M. (ed.) (1984). *Antarctic ecology*, Vols 1 and 2. Academic Press, London.

Laws, R. (1989). *Antarctica: the last frontier.* Boxtree, London.

Laws, R. M. and Franks, F. (ed.) (1990). Life at low temperatures. *Philosophical Transactions of the Royal Society of London*, **B326**, 515–697.

Longton, R. E. (1988). *Biology of polar bryophytes and lichens.* Cambridge University Press.

Lorius, C. (1991). *Glaces de l'Antarctique: une mémoire, des passions.* Éditions Odile Jacob, Paris.

Melnikov, I. A. (1997). *The Arctic sea ice ecosystem.* Gordon & Breach, Amsterdam.

Pielou, E. C. (1994). *A naturalist's guide to the Arctic.* University of Chicago Press.

Rey, L. (1984). *The challenging and elusive Arctic regions.* Significant Issue Series, Center for Strategic and International Studies, Georgetown University, **6** (5).

Reynolds, J. F. and Tenhunen, J. D. (1996). *Landscape function and disturbance in Arctic tundra.* Springer, Berlin.

Sage, B. (1986). *The Arctic and its wildlife.* Croom Helm, London.

Scoresby, W. (1820). *An account of the Arctic regions with a history and description of the northern whale-fishery.* Vol. 1, *The Arctic,* Vol. 2, *The whale fishery.* Archibald Constable, Edinburgh. (Reprinted 1969 by David & Charles, Newton Abbot)

Selkirk, P. M., Seppelt, R. D., and Selkirk, D. R. (1990). *Subantarctic Macquarie Island: environment and biology.* Cambridge University Press.

Smith, W. O., Jr. (ed.) (1990). *Polar oceanography.* Part A, *Physical science.* Part B, *Chemistry, biology, and geology.* Academic Press, San Diego.

Stonehouse, B. (1989). *Polar ecology.* Blackie, Glasgow.

Stonehouse, B. (1990). *North pole south pole: a guide to the ecology and resources of the Arctic and Antarctic.* Prion, London.

Sugden, D. (1982). *Arctic and Antarctic: a modern geographical synthesis.* Blackwell, Oxford.

Vincent, W. F. (1988). *Microbial ecosystems of Antarctica.* Cambridge University Press.

Wadhams, P., Dowdswell, J. A., and Schofield, A. N. (ed.) (1995). The Arctic and environmental change. *Philosophical Transactions of the Royal Society of London,* **A352**, 197–385.

Walton, D. W. H. (ed.) (1987). *Antarctic science.* Cambridge University Press.

Woodin, S. J. and Marquiss, M. (ed.) (1997). *Ecology of Arctic environments.* Blackwell Science.

Zenkevitch, L. (1963). *Biology of the seas of the U.S.S.R.* Unwin, London.

Selected references

It has not been possible to give references for everything mentioned in the text but it is hoped that the further reading listed above and the following selected references will enable the reader to locate most of the sources used. For a chapter in a multi-author work the names of the author(s) and editor(s) (e.g. Heywood in Laws 1984) are given in the text and the reference appears in this list only under the name(s) of the editor(s).

Aleksandrova, V. D. (1980). *The Arctic and Antarctic: their division into geobotanical areas*, (trans. D. Love). Cambridge University Press.

Ancel, A., Visser, H., Handrich, Y., Masman, D., and Le Maho, Y. (1997). Energy saving in huddling penguins. *Nature*, **385**, 304–5.

Arnaud, P. M. (1974). Contribution à la bionomie marine benthique des régions antarctiques et subantarctiques. *Téthys*, **6** (3), 465–656.

Arntz, W. E., Brey, T., and Gallardo, V. A. (1994). Antarctic zoobenthos. *Oceanography and Marine Biology: an Annual Review*, **32**, 241–304.

Barrett, P. J. (1991). Antarctica and global climatic change: a geological perspective. In *Antarctica and global climate change*, (ed. C. Harris and B. Stonehouse), pp. 35–50. Belhaven Press, London.

Becquevort, S., Mathot, S., and Lancelot, C. (1992). Interactions in the microbial community of the marginal ice zone of the northwestern Weddell Sea through size distribution analysis. *Polar Biology*, **12**, 211–18.

Benninghoff, W. S. and Benninghoff, A. S. (1985). Wind transport of electrostatically charged particles and minute organisms in Antarctica. In *Antarctic nutrient cycles and food webs*, (ed. W. R. Siegfried, P. R. Condy, and R. M. Laws), pp. 592–6. Springer, Berlin.

Bischoff, B. and Wiencke, C. (1995). Temperature ecotypes and biogeography of Acrosiphonales (Chlorophyta) with Arctic–Antarctic disjunct and Arctic/cold-temperate distributions. *European Journal of Phycology*, **30**, 19–27.

Block, W. (1994). Terrestrial ecosystems: Antarctica. *Polar Biology*, **14**, 293–300.

Block, W., Burn, A. J., and Richard, K. J. (1984). An insect introduction to the maritime Antarctic. *Biological Journal of the Linnean Society*, **23**, 33–9.

Broady, P., Given, D., Greenfield, L., and Thompson, K. (1987). The biota and environment of fumaroles on Mt Melbourne, Northern Victoria Land. *Polar Biology*, **7**, 97–113.

Brouwer, P. E. M. (1996). *In situ* photosynthesis and estimated annual production of the red macroalga *Myriogramme mangini* in relation to underwater irradiance at Signy Island (Antarctica). *Antarctic Science*, **8**, 245–52.

Burckle, L. H. and Pokras, E. M. (1991). Implications of a Pliocene stand of

Nothofagus (southern beech) within 500 kilometres of the South Pole. *Antarctic Science*, **3**, 389–403.

Burger, A. E. (1985). Terrestrial food webs in the sub-Antarctic: island effects. In *Antarctic nutrient cycles and food webs*, (ed. W. R. Siegfried, P. R. Condy, and R. M. Laws), pp. 582–96. Springer, Berlin.

Cameron, R. E. (1969). Abundance of microflora in soils of desert regions. *Jet Propulsion Laboratory, Technical Report*, 32-1378, Pasadena, CA.

Cameron, R. E., King, J., and David, C. N. (1970). Microbiology, ecology and microclimatology of soil sites in dry valleys of southern Victoria Land, Antarctica. In *Antarctic ecology*, (ed. M. W. Holdgate), pp. 702–16. Academic Press, London.

Campbell, I. B. and Claridge, G. G. C. (1987). *Antarctica: soils, weathering processes and environment.* Elsevier, Amsterdam.

Chapman, V. J. (ed.) (1977). *Wet coastal ecosystems.* Elsevier, Amsterdam.

Christensen, T. (1991). Arctic and sub-Arctic soil emissions: possible implications for global climate change. *Polar Record*, **27**, 205–10.

Clarke, A. (1993). Life in cold water: the physiological ecology of polar marine ectotherms. *Oceanography and Marine Biology: an Annual Review*, **21**, 341–453.

Clarke, A. and Leakey, R. J. G. (1996). The seasonal cycle of phytoplankton, macronutrients and the microbial community in a nearshore Antarctic marine ecosystem. *Limnology and Oceanography*, **41**, 1281–99.

Clymo, R. S. and Pearce, D. M. E. (1995). Methane and carbon dioxide production in, transport through, and efflux from a peatland. *Philosophical Transactions of the Royal Society of London*, **A350**, 249–59.

Cocks, L. R. M. (ed.) (1981). *The evolving earth.* British Museum (Natural History) and Cambridge University Press.

Cook, G. (ed.) (1990). *The future of Antarctica: exploitation versus preservation.* Manchester University Press.

Coope, G. R. (1986). The invasion and colonization of the North Atlantic islands: a palaeoecological solution to a biogeographic problem. *Philosophical Transactions of the Royal Society of London*, **B314**, 619–35.

Craig, P. C. and McCart, P. J. (1975). Classification of stream types in Beaufort Sea drainages between Prudhoe Bay and the Mackenzie Delta, N.W.T., Canada. *Arctic and Alpine Research*, **7**, 183–98.

Cripps, G. C. and Priddle, J. (1991). Hydrocarbons in the Antarctic marine environment. *Antarctic Research*, **3**, 233–50.

Davidson, M. M. and Broady, P. A. (1996). Analysis of gut contents of *Gomphiocephalus hodgsoni* Carpenter (Collembola: Hypogastruridae) at Cape Geology, Antarctica. *Polar Biology*, **16**, 463–7.

Davis, N. (1996). The Arctic wasteland: a perspective on Arctic pollution. *Polar Record*, **32**, 237–248.

Davis, R. C. (1980). Peat respiration and decomposition in Antarctic terrestrial moss communities. *Biological Journal of the Linnean Society*, **14**, 39–49.

Davis, R. C. (1981). Structure and function of two Antarctic terrestrial moss communities. *Ecological Monographs*, **51**, 125–43.

Dayton, P. K., Robilliard, G. A., and Paine, R. T. (1970). Benthic faunal zonation as a result of anchor ice at McMurdo Sound, Antarctica. In *Antarctic ecology*, Vol. 1, (ed. M. W. Holdgate), pp. 244–58. Academic Press, London.

De Baar, H. J. W., De Jong, J. T. M., Bakker, D. C. E., Löscher, B. M., Veth, C., Bathmann, U. *et al.* (1995). Importance of iron for plankton blooms and carbon dioxide drawdown in the Southern Ocean. *Nature*, **373**, 412–15.

De Freitas, C. R. and Symon, L. V. (1987). A bioclimatic index of human survival times in the Antarctic. *Polar Record*, **23**, 651–9.

Dell, R. K. (1972). Antarctic benthos. *Advances in Marine Biology*, **10**, 1–216.

De Long, E. F., Wu, K. Y., Prézelin, B. B., and Jovine, R. V. M. (1994). High abundance of Archaea in Antarctic marine picoplankton. *Nature*, **371**, 695–7.

Dickman, M. and Ouellet, M. (1987). Limnology of Garrow Lake, NWT, Canada. *Polar Record*, **23**, 531–49.

Dring, M. J. (1992). *The biology of marine plants*. Cambridge University Press.

Dunton, K. H. (1992). Arctic biogeography: the paradox of the marine benthic fauna and flora. *Trends in Ecological Evolution*, **7**, 183–9.

Dunton, K. and Schell, D. M. (1987). Dependence of consumers on macroalgal (*Laminaria solidungula*) carbon in an Arctic kelp community: ^{13}C evidence. *Marine Biology (Berlin)*, **93**, 615–25.

Ellis-Evans, J. C. (1982). Seasonal microbial activity in Antarctic freshwater lake sediments. *Polar Biology*, **1**, 129–40.

Ellis-Evans, J. C. and Wynn-Williams, D. (1996). A great lake under the ice. *Nature*, **381**, 644–6.

El-Sayed, S. Z. (ed.) (1994). *Southern Ocean ecology: the BIOMASS perspective*. Cambridge University Press.

Enzenbacher, D. J. (1994). Antarctic tourism: an overview of 1992/93 season activity, recent developments, and emerging issues. *Polar Record*, **30**, 105–16.

European Commission (1995). *Global change and Arctic terrestrial ecosystems*. Ecosystems Research Report 10 (EUR 15519 EN), Brussels.

Ewing, M. and Donn, W. L. (1956). A theory of ice ages. *Science*, **123**, 1061–6.

Falk, K. and Møller, S. (1995). Satellite tracking of high-arctic northern fulmars. *Polar Biology*, **15**, 495–502.

Fitzhugh, W. W. and Kaplan, S. A. (1982). *Inua: spirit world of the Bering Sea eskimo*. Smithsonian Institution Press, Washington DC.

Fogg, G. E. (1967). Observations on the snow algae of the South Orkney Islands. *Philosophical Transactions of the Royal Society of London*, **B252**, 279–87.

Fogg, G. E. (1991*a*). The phytoplanktonic ways of life. *New Phytologist*, **118**, 191–232.

Fogg, G. E. (1991*b*). Changing productivity of the oceans in response to a changing climate. *Annals of Botany*, **67**(suppl. 1), 57–60.

Fogg, G. E. (1992). *A history of Antarctic science*. Cambridge University Press.

Fogg, G. E. and Thake, B. (1987). *Algal cultures and phytoplankton ecology*, (3rd edn). University of Wisconsin Press.

Friedmann, E. I. (ed.) (1993). *Antarctic microbiology*. Wiley–Liss, New York.

Friedmann, E. I., McKay, Ch. P., and Nienow, J. A. (1987). The cryptoendolithic microbial environment in the Ross Desert of Antarctica: satellite-transmitted continuous nanoclimate data 1984 to 1986. *Polar Biology*, **7**, 273–87.

Garrison, D. L. and Buck, K. R. (1986). Organism losses during ice melting: a serious bias in sea ice community studies. *Polar Biology*, **6**, 237–9.

Goldman, C. R., Mason, D. T., and Wood, B. J. B. (1963). Light injury and inhibition in Antarctic freshwater phytoplankton. *Limnology and Oceanography*, **8**, 313–22.

Green, W. J. and Friedmann, E. I. (ed.) (1993). *Physical and biogeochemical processes in Antarctic lakes*. Antarctic Research Series, Vol. 59, American Geophysical Union, Washington DC.

Grinde, B. (1983). Vertical distribution of the snow alga *Chlamydomonas nivalis* (Chlorophyta, Volvocales). *Polar Biology*, **2**, 159–62.

Guillard, R. R. L. and Kilham, P. (1977). The ecology of marine plankton diatoms. In *The biology of diatoms*, (ed. D. Werner), pp. 372–469. Blackwell, Oxford.

Gutt, J. (1995). The occurrence of sub-ice algal aggregations off northeast Greenland. *Polar Biology*, **15**, 247–52.

Hansen, K. (1967). The general limnology of Arctic lakes as illustrated by examples from Greenland. *Meddelelser om Grønland*, **178** (3), 1–77.

Harris, C. M. (1991). Environmental effects of human activities on King George Island, South Shetland Islands, Antarctica. *Polar Record*, **27**, 193–204.

Hawes, I. and Brazier, P. (1991). Freshwater stream ecosystems of James Ross Island, Antarctica. *Antarctic Science*, **3**, 265–71.

Hawes, I., Howard-Williams, C., and Vincent, W. F. (1992). Desiccation and recovery of antarctic cyanobacterial mats. *Polar Biology*, **12**, 587–94.

Hayes, P. K., Whitaker, T. M., and Fogg, G. E. (1984). The distribution and nutrient status of phytoplankton in the Southern Ocean between 20° and 70°W. *Polar Biology*, **3**, 153–65.

Headland, R. (1984). *The Island of South Georgia*. Cambridge University Press.

Hegseth, E. N. (1992). Sub-ice algal assemblages of the Barents Sea: species composition, chemical composition, and growth rates. *Polar Biology*, **12**, 485–96.

Heywood, R. B. (1977). A limnological survey of the Ablation Point area, Alexander Island, Antarctica. *Philosophical Transactions of the Royal Society of London*, **B279**, 39–54.

Hobbie, J. E. (1984). Polar limnology. In *Lakes and reservoirs*, (ed. F. S. Taub), pp. 63–105. Elsevier, Amsterdam.

Horner, R. A. (ed.) (1985). *Sea ice biota*. CRC Press, Boca Raton, FL.

Horner, R., Ackley, S. F., Dieckmann, G. S., Gulliksen, B., Hoshiai, T., Legendre, L. *et al.* (1992). Ecology of sea ice biota: 1. Habitat, terminology, and methodology. *Polar Biology*, **12**, 417–27.

Huntley, B. (1991). How plants respond to climate change: migration rates, individualism and the consequences for plant communities. *Annals of Botany*, **67**(Suppl. 1), 15–22.

Huntley, M. E., Lopez, M. D. G., and Karl, D. M. (1991). Top predators in the Southern Ocean: a major leak in the biological carbon pump. *Science*, **253**, 64–6.

Hutchinson, G. E. (1957). *A treatise on limnology*, Vol. 1. Wiley, New York.

James, M. R., Pridmore, R. D., and Cummings, V. J. (1995). Plankton communities of melt ponds on the McMurdo Ice Shelf, Antarctica. *Polar Biology*, **15**, 555–67.

Jónasson, P. M. (ed.) (1979). Ecology of eutrophic, subarctic Lake Mývatn and the River Laxa. *Oikos*, **32**, 1–308.

Jónasson, P. M. (ed.) (1992). Ecology of oligotrophic, subarctic Thingvallavatn. *Oikos*, **64**, 1–437.

Jones, A. E. and Shanklin, J. D. (1995). Continued decline of total ozone over Halley, Antarctica, since 1985. *Nature*, **376**, 409–11.

Kain, J. M. (1989). The seasons in the subtidal. *British Phycological Journal*, **24**, 203–15.

Kalff, J. and Welch, H. E. (1974). Phytoplankton production in Char Lake, a natural polar lake, and Meretta Lake, a polluted polar lake, Cornwallis Island, Northwest Territories. *Journal of the Fisheries Research Board of Canada*, **31**, 621–36.

Kallio, P. and Valanne, N. (1975). On the effects of continuous light on photosynthesis in mosses. In *Fennoscandian tundra ecosystems*. Part 1, *Plants and microorganisms*, (ed. F. E. Wielgolaski), pp. 149–62, Springer, New York.

Kappen, L. and Straka, H. (1988). Pollen and spores transport into the Antarctic. *Polar Biology*, **8**, 173–80.

Karentz, D. (1991). Ecological considerations of Antarctic ozone depletion. *Antarctic Science*, **3**, 3–11.

Kerry, K. R. and Hempel, G. (ed.) (1990). *Antarctic ecosystems. Ecological change and conservation.* Springer, Berlin.

King, J. E. (1983). *Seals of the world*, (2nd edn). British Museum (Natural History) and Oxford University Press.

Kirst, G. O. and Wiencke, C. (1995). Ecophysiology of polar algae. *Journal of Phycology*, **31**, 181–99.

Kumar, N., Anderson, R. F., Mortlock, R. A., Froelich, P. N., Kubik, P., Dittrich-Hannen, B. *et al.* (1995). Increased biological productivity and export production in the glacial Southern Ocean. *Nature*, **378**, 675–80.

Laws, R. M. (1977). The significance of vertebrates in the Antarctic marine ecosystem. In *Adaptations within Antarctic ecosystems*, ed. G. A. Llano, pp. 411–38. Smithsonian Institution, Washington DC.

Leader-Williams, N. (1985). The sub-Antarctic islands – introduced species. In *Key environments: Antarctica*, (ed. W. N. Bonner and D. W. H. Walton), pp. 318–28. Pergamon, Oxford.

Legendre, L., Ackley, S. F., Dieckmann, G. S., Gulliksen, B., Horner, R., Hoshiai, T. *et al.* (1992). Ecology of sea ice biota. 2. Global significance. *Polar Biology*, **12**, 429–44.

Levere, T. H. (1993). *Science and the Canadian Arctic: a century of exploration, 1818–1918.* Cambridge University Press.

Livingstone, D. A. (1963). Alaska, Yukon, Northwest Territories, and Greenland. In *Limnology in North America*, (ed. D. G. Frey), pp. 559–74. University of Wisconsin Press.

Lizotte, M. P., Sharp, T. R., and Priscu, J. C. (1996). Phytoplankton dynamics in the stratified water column of Lake Bonney, Antarctica. 1. Biomass and productivity during the winter-spring transition. *Polar Biology*, **16**, 155–62.

Lizotte, M. P. and Sullivan, C. W. (1992). Photosynthetic capacity in microalgae associated with Antarctic pack ice. *Polar Biology*, **12**, 497–502.

Lønne, O. J. and Gulliksen, B. (1991). On the distribution of sympagic macro-fauna in the seasonally ice covered Barents Sea. *Polar Biology*, **11**, 457–69.

Marshall, W. A. (1996). Biological particles over Antarctica. *Nature*, **383**, 680.

Mathot, S., Dandois, J.-M., and Lancelot, C. (1992). Gross and net primary production in the Scotia–Weddell Sea sector of the Southern Ocean during spring 1988. *Polar Biology*, **12**, 321–32.

Matthews, J. A. (1992). *The ecology of recently-deglaciated terrain.* Cambridge University Press.

May, R. M., Beddington, J. R., Clark, C. W., Holt, S. J., and Laws, R. M. (1979). Management of multispecies fisheries. *Science*, **205**, 267–77.

Medlin, L. K., Lange, M., and Baumann, M. E. M. (1994). Genetic differentiation among three colony-forming species of *Phaeocystis*: further evidence for the phylogeny of the Prymnesiophyta. *Phycologia*, **33**, 199–212.

Miles, P. and Wright, N. J. R. (1978). An outline of mineral extraction in the Arctic. *Polar Record*, **19**, 11–38.

Miller, D. G. M. and Hampton, I. (1989). *Biology and ecology of the Antarctic krill (Euphausia superba* Dana): *a review.* BIOMASS Scientific Series no. 9. SCAR and SCOR, Scott Polar Research Institute, Cambridge.

Murphy, E. J., Clarke, A., Symon, C., and Priddle, J. (1995). Temporal variation in Antarctic sea-ice: analysis of a long-term fast-ice record from the South Orkney Islands. *Deep-Sea Research*, **42**, 1045–62.

Paine, R. (1988). Reindeer and caribou *Rangifer tarandus* in the wild and under pastoralism. *Polar Record*, **24**, 31–42.

Pisek, A. (1960). Pflanzen der Arktis und des Hochgebirges. In *Encyclopedia of plant physiology*, Vol. V(2), (ed. A. Pirson), pp. 376–414. Springer, Berlin.

Priddle, J. and Heywood, R. B. (1980). Evolution of Antarctic lake ecosystems. *Biological Journal of the Linnean Society*, **14**, 51–66.

Raven, J. A. (1984). *Energetics and transport in aquatic plants.* Liss, New York.

Rees, W. G. (1993). A new wind-chill nomogram. *Polar Record*, **29**, 229–34.

Reynolds, C. S. (1992). The role of fluid motion in the dynamics of phytoplankton in lakes and rivers. In *Aquatic ecology: scale, pattern and process*, (ed. P. S. Giller, A. G. Hildrew, and D. G. Raffaelli), pp. 141–87. Blackwell, Oxford.

Rigler, F. H. (1978). Limnology in the high Arctic: a case study of Char Lake. *Verhandlungen der Internationale Vereinigung für Limnologie*, **20**, 127–40.

Robinson, C., Hill, H. J., Archer, S., Leakey, R. J. G., Boyd, P. W., and Bury, S. J. (1995). Scientific diving under sea ice in the Southern Ocean. *Underwater Technology*, **21**, 21–7.

Robinson, D. H., Arrigo, K. R., Iturriaga, R., and Sullivan, C. W. (1995). Microalgal light-harvesting in extreme low-light environments in McMurdo Sound, Antarctica. *Journal of Phycology*, **31**, 508–20.

Rodhe, W. (1955). Can plankton production proceed during winter darkness in sub-arctic lakes? *Verhandlungen der Internationale Vereinigung für Limnologie*, **12**, 117–22.

Røen, U. (1994). A theory for the origin of the Arctic freshwater fauna. *Verhandlungen der Internationale Vereinigung für Limnologie*, **25**, 2409–12.

Rosswall, T. and Heal, O. W. (ed.) (1975). *Structure and function of tundra ecosystems.* Ecological Bulletin No. 20, Swedish Natural Science Council, Stockholm.

Ryan, P. G. and Watkins, B. P. (1989). The influence of physical factors and ornithogenic products on plant and arthropod abundance at an inland nunatak group in Antarctica. *Polar Biology*, **10**, 151–60.

Sambrotto, R. N., Goering, J. J., and McRoy, C. P. (1984). Large yearly production of phytoplankton in the western Bering Sea. *Science*, **225**, 1147–50.

Schlichting, H. E., Jr., Speziale, B. J., and Zink, R. M. (1978). Dispersal of algae and protozoa by antarctic flying birds. *Antarctic Journal of the U.S.*, **13**, 147–9.

Schreiber, A., Eisinger, M., and Storch, V. (1996). Allozymes characterize sibling species of bipolar Priapulida (*Priapulis*, *Priapulopsis*). *Polar Biology*, **16**, 521–6.

Seligman, G. (1980). *Snow structure and ski fields*, (3rd edn). International Glaciological Society, Cambridge.

Sheldon, J. F. (1988). Oil versus caribou in the Arctic: the great debate. *Polar Record*, **24**, 95–100.

Shreeve, R. S. and Peck, L. S. (1995). Distribution of pelagic larvae of benthic marine invertebrates in the Bellingshausen Sea. *Polar Biology*, **15**, 369–74.

Smith, R. C., Prézelin, B. B., Baker, K. S., Bidigare, R. R., Boucher, N. P., Coley, T. *et al.* (1992). Ozone depletion: ultraviolet radiation and phytoplankton biology in Antarctic waters. *Science*, **255**, 952–9.

Smith, R. I. L. (1972). *Vegetation of the South Orkney Islands with particular reference to Signy Island.* British Antarctic Survey, Scientific Report No. 68, London.

Smith, R. I. L. (1984). Colonization and recovery by cryptogams following recent volcanic activity on Deception Island, South Shetland Islands. *British Antarctic Survey Bulletin*, No. 62, 25–51.

Smith, R. I. L. (1985). *Nothofagus* and other trees stranded on islands in the Atlantic sector of the Southern Ocean. *British Antarctic Survey Bulletin*, No. 66, 47–55.

Stirling, I. and Calvert, W. (1983). Environmental threats to marine mammals in the Canadian Arctic. *Polar Record*, **21** 433–49.

Stone, R. (1995). Russian Arctic battles pipeline leak. *Science*, **268**, 796–7.

Stonehouse, B. (1992). Monitoring shipborne visitors in Antarctica: a preliminary field study. *Polar Record*, **28**, 213–8.

Tedrow, J. C. F. (1977). *Soils of the polar landscapes*. Rutgers University Press, New Brunswick.

Thomas, W. H. and Duval, B. (1995). Sierra Nevada, California. U.S.A., snow algae: snow albedo changes, algal-bacterial interrelationships, and ultraviolet radiation effects. *Arctic and Alpine Research*, **27**, 389–99.

Tilzer, M., Elbrächter, M., Gieskes, W. W., and Beese, B. (1986). Light temperature interactions in the control of photosynthesis in Antarctic phytoplankton. *Polar Biology*, **5**, 105–11.

Triggs, G. D. (ed.) (1987). *The Antarctic Treaty regime*. Cambridge University Press.

Vincent, W. F. (1987). Antarctic limnology. In *Inland waters of New Zealand*, (ed. A. B. Viner), pp. 379–412. SIPC, New Zealand.

Vincent, W. F. and Vincent, C. L. (1982). Factors controlling phytoplankton production in Lake Vanda (77°S). *Canadian Journal of Fisheries and Aquatic Science*, **39**, 1602–9.

Vincent, W. F., Castenholz, R. W., Downes, M. T., and Howard-Williams, C. (1993). Antarctic cyanobacteria: light, nutrients, and photosynthesis in the microbial mat environment. *Journal of Phycology*, **29**, 745–55.

Vitebsky, P. (1989). Reindeer herders of northern Yakutia: a report from the field. *Polar Record*, **25**, 213–18.

Volkov, A. E. and De Korte, J. (1994). Protected nature areas in the Russian Arctic. *Polar Record*, **30**, 299–310.

Walton, D. W. H. (1982). Instruments for measuring biological microclimates for terrestrial habitats in polar and high alpine regions: a review. *Arctic and Alpine Research*, **14**, 275–86.

Wand, U., Schwarz, G., Brüggemann, E., and Braüer, K. (1997). Evidence for physical and chemical stratification in Lake Untersee (central Dronning Maud Land, East Antarctica). *Antarctic Science*, **9**, 43–5.

Weissenberger, J., Dieckmann, G., Gradinger, R., and Spindler, M. (1992). Sea ice: a new technique to analyze and display the interstitial environment. *Limnology and Oceanography*, **37**, 179–83.

Weller, G. and Holmgren, B. (1974). The microclimate of the arctic tundra. *Journal of Applied Meteorology*, **13**, 854–62.

Whitaker, T. M. (1977). Sea ice habitats of Signy Island (South Orkneys) and their primary productivity. In *Adaptations within Antarctic ecosystems*, (ed. G. A. Llano), pp. 75–82. Smithsonian Institution, Washington DC.

Wielgolaski, F. E. (1975). Primary production of tundra. In *Photosynthesis and productivity in different environments*, (ed. J. P. Cooper), pp. 75–106. Cambridge University Press.

Wiencke, C. (1996). Recent advances in the investigation of Antarctic macroalgae. *Polar Biology*, **16**, 231–40.

Williams, T. D. (1995). *The penguins*. Oxford University Press.

Wolff, E. W. (1990). Signals of atmospheric pollution in polar snow and ice. *Antarctic Science*, **2**, 189–205.

Wortmann, H. (1995). Medizinische Untersuchungen zur Circadian-rhythmik und zum Verhalten bei Überwinterern auf einer antarktischen Forschungsstation. *Berichte zur Polarforschung*, **169**, 1–261.

Index

N.B. The 'letter-by-letter' system has been used. *Italic* numbers denote reference to illustrations.